Bus & Coach Recognition

VETERAN & VINTAGE

G. B. WISE

LONDON

IAN ALLAN LTD

First published 1989

ISBN 0 7110 1858 8

Published by Ian Allan Ltd, Shepperton, Surrey; and printed by Ian Allan Printing Ltd at their works at Coombelands in Runnymede, England

All photographs are by the author or from his collection unless credited otherwise.

Front cover:
Built in 1948, this Daimler CVD6 with Burlingham bodywork was photographed at the North Weald Rally on 19 June 1988. Kevin Lane

Back cover:
AEC Regent HUM 401, seen during the Pennine Rally in July 1988.
John Robinson

Above:
The AEC Regent was the great rival of the Leyland Titan and in 1950-52 London Transport purchased 76 Regent IIIs with lowbridge 53-seat Weymann bodies, 11 of which are now preserved. Two of these (RLH 23 and 48) were at the Wendover rally in July 1986, one of the smaller but very popular bus rallies held each year.

Contents

Introduction

This book seeks to give an informative overview of the development of the British bus and coach and to put into their historical context the many vehicles which are now preserved. It sets out to do this by providing basic technical information, recognition features and general information on each chassis manufacturer and to illustrate the most common or important types. In this way any chassis type can be easily referred to and its main features ascertained. Only British vehicles have been included although there are a number from other parts of the world preserved in these islands.

The British bus and coach scene has undergone enormous changes over the past 80 years and those who witnessed the industry's beginnings would undoubtedly be amazed at today's vehicles and operations. The early years were characterised by unreliability and underpowered vehicles, of small local body-builders developing their own designs from earlier experience with horse-drawn carts and buses. They were also times of uncertainty when a service introduced with hope and optimism often lasted only a short time and was soon forgotten. From World War 1 onwards rapid development took place and, with companies like AEC and Leyland giving a clear lead, many smaller manufacturers quickly modernised their designs and introduced new ones aimed at specific sectors of the market. Variety has been the hallmark of British bus and coach operations through the years and is, almost surprisingly, no less so today. This book portrays something of the development and variety of the vehicles which have appeared and of which, fortunately for us, so many have survived into the present day.

The bus and coach preservation movement is growing steadily as more vehicles are retired and bought by enthusiasts. The number of rallies at which they may be observed is increasing as well and whilst many of the vehicles exhibited are modern there are always many older examples to admire. In recent years a number of museums devoted to the subject have been established and in these can be found many non-operational vehicles on which restoration is taking place as well as those allowed out to special events. The rally calendar has now grown to the extent that on most summer weekends at least one will be taking place. Many of these are small one-day affairs but others, of one or two days duration, attract a large number of entrants. The best source of information about these is in *Buses* which features a regular article on the preservation scene, lists museums and rallies in its April issue each year, and contains information on special events in its advertisement section. There are growing numbers of books on buses, many from small specialised publishers covering fleet histories and vehicle builders, and one to be recommended is D. J. Slater's *Preserved Buses, Trolleys and Trams Stock Book* together with its supplement. Although somewhat out of date it lists most preserved buses and coaches and is an invaluable basic sourcebook. K. A. Jenkinson's *Preserved Buses* (now out of print) is also useful.

I hope this book will be an invaluable guide to the identification of earlier generations of bus and coach design, and that it will answer many questions as to the background and technical details of what is on view at the various rallies and museums. I am indebted to all those who have helped me in its production, not least Stephen Morris, whose knowledge of the subject has been invaluable, and my wife, Maureen, for her constant encouragement.

Graham Wise
Great Missenden
July 1988

Note: *All engine capacities, unless otherwise specified, are in litres.*

ADC Associated Daimler Co Ltd

Built: 1926-28

Bodywork types: Single- and double-deck bus and coach

Engines: AEC – 4-cyl 5.1 petrol
Daimler – 6-cyl 3.568 petrol

Wheelbase lengths: 14ft 6in to 16ft 0in

Typical bodywork makes:
Dodson, Hall Lewis, LGOC, Phoenix, Strachan & Brown

Recognition features:
Wide, squat radiators with detail differences as applied to each type, and carrying Associated-Daimler, AEC or LGOC names on a plate affixed to the radiator top tank. The LGOC double-deck vehicles had no windscreens due to Metropolitan Police regulations and the single-deckers were a mix of normal and forward control types. Many earlier vehicles fitted with solid tyres.

General remarks:
The Associated Daimler Co was formed in 1926 by AEC and Daimler as a joint marketing venture using the AEC 4- and 5-type chassis and either AEC or Daimler engines. Chassis designations were suffixed A or D to indicate which make of engine was installed. The first model was the straight-frame 413 which had its origins in the AEC 411 of 1923 but was significantly improved by such things as hand and foot brakes operating on the rear brake drums and pneumatic tyres. A longer wheelbase version was designated 415. The 416 was a forward control drop-frame chassis and the 417 its normal control equivalent. The old AEC 405 (the LGOC NS type) was developed as the 422 and as a three-axle chassis, the 802 (the LGOC LS type). AEC trolleybuses were also marketed under the ADC name.

Below:
RR 7635 was a fairly typical ADC 416A forward control chassis fitted with Hall Lewis dual doorway bus bodywork. Belonging to West Bridgford UDC it was photographed near the depot in April 1934. Common features of the period are the squat radiator, crank-handle and exterior mounted horn. The Associated Daimler name appears on the radiator toptank. G. H. F. Atkins

In 1927 the Daimler-engined normal control 423 and the forward control 424 were introduced followed by the AEC-engined forward control 426 and normal control 427 which replaced the earlier 416 and 417. The 425, a modified 426, was supplied only to United Automobile Services. The final design was the 427 of which only two were built in 1928 and fitted with small 18-seat coach bodies. The designation 419 was also used for a chassis supplied to the LGOC and fitted with 28-seat all-weather coach bodies and given the LGOC type code AW. In July 1928 the decision was taken not to continue the Association and from then on AEC and Daimler proceeded to market their products separately.

Associated Equipment Company London General Omnibus Company

Built: 1909-26

Bodywork types: Double- and single-deck bus, charabanc

Engines: LGOC – 4-cyl 28hp petrol (X, early B, K types)
AEC – 4-cyl 35hp petrol (S, NS types)
4-cyl 40hp petrol (later B type)

Wheelbase lengths: 12ft 10⅝in (B type)
13ft 0in (B type)
14ft 0in (B type)
14ft 2½in (K type — AEC Models 301-5)
14ft 6in (Single-deck B type, AEC Model 411)
14ft 11in (S type — AEC Model 401, Models 403, 502/7)
15ft 6in (NS type — AEC Models 405/7/8/10/22)

Typical bodywork makes:
Vanguard, Dodson (LGOC vehicles), Brush, Ransomes, Strachan & Brown, United Automobile Services (provincial vehicles)

Recognition features:
X and B types are normal control, the remainder forward control. Wide, squat radiators with top tank moulding incorporating the filler-cap and bearing the LGOC bullseye and bar — LGOC examples carrying the word General and provincials AEC. Early examples

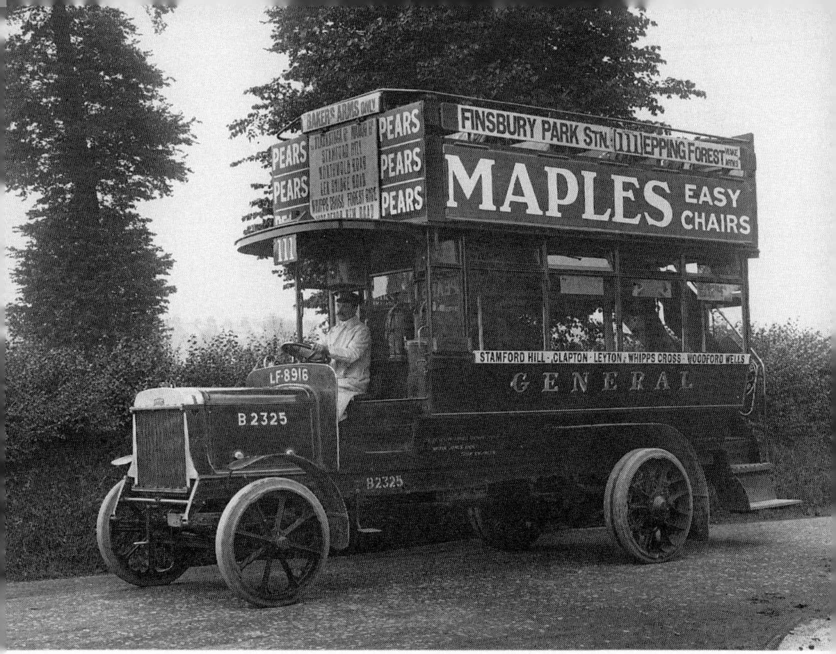

Below left:
Many early chassis were used for goods or passenger work — in some cases both — and had very high floor levels. Solid tyres were standard and bodywork was often locally built. This AEC Y type dating from around 1917 was photographed before leaving the Walthamstow factory. A 3 to 4 ton model it had a Tylor 45bhp engine to enable some standardisation of parts used in civil and military vehicles.

Above:
Among the most famous of London's buses is the AEC B-type of which over 3,200 were built between 1910 and 1919, and which the War Office commandeered in considerable numbers. Solid tyres and the lack of a windscreen to protect the driver conformed to Metropolitan Police regulations. The bodywork displays its horsebus ancestry with 16 inside and 18 outside seats reached by the outside staircase. The radiator of B2325 is marked LGOC. AEC

— and all LGOC — have solid tyres. Safety rails between front and rear axles. London double-deck examples are all opentop, without windscreens and with solid tyres, though some were later fitted with pneumatics.

General remarks:

The origins of AEC can be traced to Jacques Lafitte who began horse bus operations in Paris in 1820. His company, reorganised in 1854, commenced services in London in 1856 as the Compagnie General des Omnibus de Londres with capital of £1 million. In 1859 its offices were moved to London and the enterprise became the London General Omnibus Co. In the ensuing years a large number of operators were absorbed, the most notable being the London Road Car Co and the London Motor Bus Co. The latter, trading as Vanguard, with a works at Walthamstow was taken over in 1908 and, with almost 300 vehicles, doubled the size of LGOC.

In 1909 the X type appeared and was admitted to be an amalgam of all that was best in current bus designs. It was of normal control and driven via a cone clutch through a chain and sprocket gearbox. Sixty more soon followed in 1910, in which year the

Above:
The LGOC K type was introduced in 1919 and, in spite of the restrictions imposed by the Metropolitan Police showed significant advances on the B type although it used the same engine. The driver was equipped with a rudimentary weathershield and the passengers benefited from lower step heights. The General name appears on the radiator and this became the norm until the appearance of the famous blue triangle in 1928 K502, dating from 1920, is now preserved.

---

famous B type took to the streets. This was basically similar to the X but later examples had a 40hp engine and single-plate clutch. In 1912 the LGOC renamed its bus building operations as the Associated Equipment Co and the B type was then produced as an AEC model. Some 3,200 were built of which over 1,000 were requisitioned for war service, production ending in 1919. A new vehicle emerged that year and was designated the K type (AEC chassis Models 301-5). This was a forward control design using a three-speed gearbox and, originally, solid tyres. It could carry 12 more passengers within the same overall dimensions as the B type. London had 1,156 Ks but about 30 others were sold to provincial operators, production ending in 1926. In 1920 the maximum permitted weight was increased to eight tons as a result of which the AEC Model 410 was introduced. Designated the S type by the LGOC it was also built for provincial operators as the 403, 502 and 507 Models. It used a multiplate clutch and three-speed gearbox and also had larger wheels than its predecessors. Over 900 were built before a larger variant, the NS, came into service in 1923 using the 403 chassis with a multidisc oil-bath clutch, four-speed constant mesh helical gearbox and double-reduction rear axle. The Associated Daimler Co took over production in 1927 and the chassis became the ADC 409 and 422, production ending in 1928. A further chassis introduced before the formation of ADC was the 411, a forward control single-decker using a cone clutch, four-speed gearbox and worm-drive rear axle.

Examples of the B, K, S and NS types have been preserved.

AEC Reliance 660

Built: 1928-32
Bodywork type: Single-deck bus and coach
Engine: AEC – 6-cyl 6.1 petrol
Wheelbase length: 15ft 6in
Typical bodywork makes:
Duple, Harrington, LGOC, Weymann

Recognition features:
Forward control design with a radiator owing much to the ADC range, having a deep, chromed surround and the first type of AEC triangle at its top (set half on the top tank and half on the radiator mesh). Large headlamps mounted each side of the radiator. Many examples had the horn mounted on the outside of the dash-panel below the windscreen.

General remarks:
The Reliance 660 was the first new chassis to appear from Southall after the ending of the ADC arrangement and under Chief Engineer G. J. Rackham. The Reliance owed much to the AEC Model 422 and ADC 426 chassis but was designed for single-deck bodywork. Using a new six-cylinder engine (AEC had used four-cylinder units up to this time) it incorporated the well-proven cone clutch and a four-speed gearbox. Although only an interim model preparing the way for the highly successful Regal and Regent chassis launched in 1929, almost 500 were constructed before production ceased in 1932. The LGOC bought a total of 48 in both bus and coach form, some coming second-hand after new building had ceased. LGOC Reliance coaches pioneered many of the successful Green Line services although they were soon replaced by Regals. All the LGOC Reliances had been sold by 1938. One East Surrey Motor Services example is known to have been fitted with a double-deck body. One Great Yarmouth registered Reliance which ended its days as a caravan is known still to exist.

Below:
The AEC Reliance 660 of 1928 was a thoroughly modern design and is typified by Speedwell Coaches' CU 9068. The Reliance name was carried on the lower radiator framing whilst the new AEC blue triangle was mounted on the toptank. The radiator itself was virtually identical to that of the old ADC 422 and 426 chassis. About two-thirds of the triangle projected below the toptank and on it appeared the LGOC bullseye with the letters AEC on the bar, a trademark that was to last the life of the company.
Ian Allan Library

AEC Regent

Built: 1929-41, 1945-69

Bodywork type: Double-deck bus

Engines:

AEC	– 6-cyl	6.1	petrol (Regent I)
	6-cyl	9.6	oil (2RT and IIIRT, Mk III)
	6-cyl	7.7	oil (Mk II)
	6-cyl	7.685	AV470 type/oil (Mk V)
	6-cyl	9.636	AV590 type oil (Mk V)
Gardner	– 6-cyl	8.4	6LW type oil (Mk V)

Wheelbase lengths:

15ft 6½in	(Regent I)
16ft 3in	(Mk II)
16ft 4in	(2RT, IIIRT, Mk III, Mk V)
18ft 7in	(Mk V option after 1956)

Typical bodywork makes:

Regent I: Brush, English Electric, MCCW, Northern Counties, Tilling, Weymann, Willowbrook

LT RT-type: Chiswick, Craven, Park Royal, Saunders, Weymann

Postwar models: Crossley, Duple, MCCW, Northern Counties, Park Royal, Roe, Weymann, Willowbrook

Variants:

Regent I	– 1929-41
Regent 2RT	– 1938-40
Regent II	– 1945-47
Regent IIIRT	– 1946-53
Regent Mk III	– 1947-57
Regent Mk IV	– 1954 (one prototype)
Regent Mk V	– 1954-69

Recognition features:

Distinctive AEC type radiator, slightly tapering in towards the base, chrome surround, slightly curved corners, central dividing strip surmounted by the AEC triangle. Headlamps mounted either side. On postwar models the radiator is deeper but of the same design. Tapered bonnet sides with mudguards extending in to meet it. Some Mk IIIs have the 'new look' front with concealed radiator covered by vertical strips and featuring the AEC triangle. The Mk Vs mostly had a concealed radiator comprising vertical slats or a black mesh screen slightly curved top and bottom, with a vertical divider surmounted by the AEC triangle, the headlamps being mounted in the front panel.

General remarks:

The Regent, introduced in 1929, was G. J. Rackham's answer to the new Leyland Titan. Its 15ft 6½in wheelbase was increased to 16ft 3in the following year and remained constant (apart from the one-inch longer RT) until the introduction of the Mk III in 1947. It used the six-cylinder engine of the Reliance together with a four-speed sliding mesh gearbox and, from 1932, a fully floating rear axle. In 1933 AEC separated from the LPTB but an agreement was reached for AEC to supply at least 90% of London Transport requirements, thus providing a secure future for the company. The 7.7-litre engine was fitted from 1934 and by the end of production in 1941 some 7,200 had been built of which 3,764 went to London Transport. In addition to these from 1938-40 some 151 Regent 2RTs were supplied to London Transport as a result of a joint AEC/LT design venture. These were the first of the famous RT-type and were powered by the 9.6-litre engine and had air-actuated four-speed pre-selector gearboxes and, unusually, air brakes. Bodywork was to London Transport's design.

Following the war, production resumed in 1945 with the Regent Mk II which was basically similar to the prewar model with a 7.7-litre engine. Production ceased in 1947 by which time almost 700 Mk IIs had been sold. The Regent Mk II was replaced by the Mk III which was available in 7ft 6in and 8ft 0in widths and powered by the 9.6-litre engine using a four-speed pre-selector gearbox and vacuum servo-assisted braking. In 1950 the chassis was lengthened to take 27ft 0in-long bodywork although the wheelbase

Top:
MV 3522, an AEC Regent demonstrator of 1932 shows clearly the radiator style which became standard from 1929. The bodywork is by Park Royal, to lowbridge height with sunken offside upper-deck gangway, half-drop windows and curved seat-tops. Note the crank-handle on the radiator front and the autovac mounted on the front bulkhead.
AEC

Above:
The Regent 2RT was a joint AEC/London Transport venture of 1938 leading to a production run of 150 vehicles. The prewar RT differed from its postwar counterpart by having sloping bottom edges to front and side cab windows and also in the larger air intake grille on the cab front. All had roof-mounted route number boxes. RT 54 and 113, seen here preserved in wartime livery, have LPTB Chiswick H30/26R bodywork.

HUDDERSFIELD
43
VIA ELLAND BRIDGE
WEBSTER'S
Yorkshire
Bitter
178
JVH 378
67
EASTERN NATIONAL

SEACROFT
25
LEEDS CITY
TRANSPORT
952
952 JUB

Left:
The AEC Regent III made its appearance in 1947 with production running for 10 years. JVH 378 was built in 1955 for Huddersfield Corporation and ended its working life with West Yorkshire PTE before being preserved. This view shows the lower postwar style radiator with the Regent name thereon, and also the larger windscreen and normal AEC style sliding driver's door. Bodywork is H61R by East Lancashire Coach Builders.

Below left:
Most AEC Regent Vs had concealed radiators with a distinctive grille and full-width front panel as seen on Leeds City Transport 952 (952 JUB) dating from 1964 and fitted with Roe rear platform bodywork. Fully restored, this bus is now in the care of the Lincolnshire Vintage Vehicle Society.

Above:
The postwar London Transport RT type was based on the Regent III chassis and went into production in 1947 with bodywork to an almost standard design from four builders. All bodies were interchangeable including those of the RTL class Leyland PD2s. An exception to the general standard design was that of the Cravens-built examples which had seven upper and five lower window-bays. Most bodies were supplied by Park Royal and Weymann and were constructed largely from steel. RT1700 (KYY 527) of 1950 carries a Weymann body of 1949 and is now preserved.

---

remained the same. In 1952 a four-speed synchromesh gearbox was fitted and production totalled over 5,900 before ending in 1957. Meanwhile the famous RT in its 3RT form had begun production in 1946. This, too, used the 9.6-litre engine but had a four-speed air-operated pre-selector gearbox and air-pressure braking. LT took 4,674 examples and others were supplied to Devon General, Rhondda, West Riding, Aberdeen, Birmingham, Coventry, Douglas, Grimsby, Halifax and St Helens. The last 3RTs were completed in 1953.

At the 1954 Commercial Motor Show AEC exhibited an underfloor-engined Regent Mk IV chassis which had been built in 1949 and fitted with a full-fronted 60-seat Crossley body. No further examples were built and the conventional layout Mk V appeared the same year. This had the largest number of options and variations of any Regent produced. Three different engines were available together with monocontrol and synchromesh gearboxes, air pressure and triple vacuum servo-assisted braking. A few were also produced with preselector gearboxes to special customer order. This led to a wide variety of design codes. Production ended in 1969 but a few were still in regular service in 1988.

AEC Regal

Built: 1929-38 and 1945-56

Bodywork types: Single-deck bus and coach

Engines:

AEC	4-cyl	5.1	petrol (Regal 4)
	4-cyl	6.6	oil (Regal 4)
	6-cyl	6.1	petrol (Regal 662)
	6-cyl	6.6	oil (Regal Mk II 0862)
	6-cyl	7.7	oil (Regal 662, Mk I, Mk III)
	6-cyl	8.8	oil (Regal 662)
	6-cyl	9.6	oil (Mk III)
	6-cyl	9.6	horizontal oil (Mk IV)
Gardner	6-cyl	8.4	6LW type oil (Regal 662)

Wheelbase lengths: 16ft 4in (Mk IV)
17ft 0in (Regal 662)
17ft 6in (Regal 662 from 1932, Regal 4, Mk I)
19ft 3in (Mk III)
20ft 0in (Mk III export model)

Typical bodywork makes:

Prewar: Brush, Burlingham, Cowieson, Duple, Harrington, Park Royal, Roe, Weymann, Willowbrook

Postwar: Burlingham, Eastern Coach Works, Harrington, Mann Egerton, Park Royal, Roe, Weymann, Willowbrook, Windover

Variants:

Regal 662	– 1929-40
Regal 4	– 1930-37
Mk II 862	– 1935-39
Mk I	– 1946-47
Mk III	– 1947-57
Mk IV	– 1950-55
Ranger 665 and Mercury 640	

Recognition features:

Distinctive AEC type radiator, slightly tapering in towards the base, chrome surround, slightly curved corners, central dividing-strip surmounted by the AEC triangle. Headlamps mounted either side. On postwar models the radiator is deeper but of the same design. Tapered bonnet sides. The Mk III has wider mudguards. The Mk IV is a full-fronted design with underfloor engine and entrance forward of the front axle. The AEC triangle normally appears on the front panel but LT examples carry the LT roundel and bar.

General remarks:

The Regal 662 was the single-deck version of the Regent using a 17ft 0in wheelbase. The 6.1-litre petrol engine was retained but a single-plate clutch was used with a sliding-mesh four-speed gearbox bolted directly to the engine. In 1931 the wheelbase was lengthened to 17ft 6in and, as with the Regent, a fully floating rear axle was fitted. The 8.8-litre oil engine was also offered, followed in 1934 by the 7.7-litre unit and in 1935 the Gardner 6LW. In 1930 a lower powered version called the Regal 4 became available. This was fitted with either the four-cylinder 5.123-litre petrol engine or an oil unit of 6.6 litres. Production continued until 1939 but only 180 examples were sold. In 1935 the 662 was redesigned to a lighter specification and became known as the Mk II. Having a shorter bonnet it allowed for an increase in seating capacity within the same overall dimensions. By the time production ceased in 1940 some 3,500 had been built.

A normal control version of the 662 was also marketed from 1930-39. This was known as the Ranger 665 or Mercury 640 depending on the engine used — the former being fitted with a six-cylinder engine and the latter one of four-cylinders. 101 examples of these were built.

The Regal Mk I was introduced in 1946 and basically followed the prewar design. There were no options available, the engine being of 7.7 litres and a friction clutch coupled to a four-speed sliding mesh gearbox plus vacuum assisted triple servo braking

Above:
The higher-set prewar-style radiator is apparent in this view of 1939 built Regal 662 AFU 927 — 1816 in the Lincolnshire Road Car fleet — on a Scunthorpe town service in the 1950s. Originally fitted with a Plaxton C35F body it was new to Enterprise (Scunthorpe) Ltd and acquired by LRCC with that undertaking in March 1950. In May of that year it was given a new Burlingham B35F body which it retained until the end of its life. Note the crank-handle shaft, the flared bodywork skirt and the emergency door immediately behind the driver's.

Below:
The AEC Ranger was the normal control version of the Regal and was notable for its long tapering bonnet and high radiator which necessitated kerbfinders on each mudguard. The fold-back roof and luggage rack are clearly visible in the view of JA 3504 of Howe's Tours at Llandudno in June 1933. G. H. F. Atkins

The London Transport RF is the most familiar version of the Regal IV and about 100 of these are now in preservation, including RF 226 (MLL 763) seen here at the 1986 Wendover Rally. All were bodied by Metro-Cammell with 37, 39 or 41 seats depending on their use as Central or Country Area buses or Green Line coaches. Although all had front entrances the Central Area buses lacked doors.

GREEN LINE
EXPRESS
724 ST. ALBANS
GREEN LINE
MLL 763

Below:
Duple coachwork without the canopy over the radiator is fitted to this 1947 Regal III which was new to Silver Service of Matlock. Happily now preserved it is seen at the 1988 North Weald Rally.

were standard. In just over a year some 1,665 were sold and it was then replaced by the Mk III early in 1948. This was the single-deck version of the 'Provincial' Regent Mk III double-decker with the 9.6-litre engine as standard along with a fluid-flywheel, four-speed pre-selector gearbox and air braking. A friction clutch and sliding mesh gearbox then became options. By the time production ceased in 1955, 3,780 had been built including over 1,600 for export.

The Regal Mk IV of 1950 saw AEC's entry into the underfloor-engined chassis market. Originally designed with a 15ft 8in wheelbase only 26 were built to this length, 16ft 4in becoming the standard. The shorter wheelbase was for bodywork of 27ft 6in × 7ft 6in whilst the longer modification allowed for 30ft 0in × 8ft 0in permitted from that year. The London Transport order for 675 was to 30ft 0in × 7ft 6in dimensions. A horizontal 9.645-litre engine was mounted between the axles on the offside and coupled to a fluid-flywheel and preselector gearbox. Air braking was standard. In 1951 a synchromesh gearbox with friction clutch became available. Production for the home market ceased in 1955 but exports continued until 1957, a total of 2,672 being built with 503 of these for export.

AEC Prewar Renown

Built: 1929-37 (Model 663), 1930-40 (Model 664)

Bodywork types: Double and single-deck bus, single-deck coach

Engines:

AEC –	6-cyl	6.1	petrol (Model 663)
	6-cyl	7.4	petrol (Models 663 and 664)
	6-cyl	7.7	oil (Models 663 and 664)
	6-cyl	8.8	oil (Model 664)
Gardner –	5-cyl	7.0	5LW type oil (Model 663)
	6-cyl	8.4	6LW type oil (Model 663)

Wheelbase lengths: 16ft 6in (Model 663), 18ft 7in (Model 664)

Typical bodywork makes:
Brush, Harrington, LGOC, Massey, MCCW, Short, Strachans (Acton) Ltd, Weymann

Recognition features:
Three-axle design, mostly double-deck. Typical AEC radiator with central divider surmounted by the AEC triangle. Upper deck bodywork was often set back behind the driving cab or of 'piano-front' design.

General remarks:
The Associated Daimler Company had supplied the London General Omnibus Company with 11 massive three-axle buses, known as the LS or 'London Six', in 1927. The Renown 663 was a development of the Regent rather than of these earlier six-wheelers, and an attempt at solving capacity problems within the maximum permitted lengths which were then 25ft 0in for a two-axle double-decker and 30ft 0in for one of three axles. The early 663 models for London were able to seat 60 passengers and had open staircases, the longer 664 was able to seat 70 with an enclosed staircase. Double-drive was standard and single tyres were fitted all round. The original power unit of the 663 was the 6.1-litre petrol engine and for the 664 the 7.4-litre unit but other options soon became available. Both had a single-plate clutch and four-speed sliding mesh gearbox but in 1932 a fluid-flywheel and four-speed preselector gearbox became an option. A total of 1,250 663s was built, all double-deckers except the final 24 which were LTC type coaches for London Transport, and 349 664s which were a mix of double and single-deck buses and single-deck coaches. Of these 48 664s were exported. Two double-deck 663s and one double and one single-deck bus 664s are now preserved.

Below:
The prewar AEC Renown is instantly recognisable by its radiator and twin rear axles which enabled 30ft-long bodywork and increased seating. Leicester CT329 (CBC 921) had 64 seats in a Northern Counties body. In prewar fashion the radiator is high-set and the windscreen small. No 329 was one of the last 663 type to be built. M. Fowler

AEC Q

Built: 1932-37

Bodywork types: Double and single-deck bus, single-deck coach

Engines: AEC – 6-cyl 7.4 petrol
6-cyl 7.73 oil

Wheelbase length: 18ft 6in

Typical bodywork makes:
Birmingham RC & W, Cravens, Duple, MCCW, Park Royal, Roe, Willowbrook

Recognition features:
Full-fronted forward control design with significant front overhang (longer on double-deck models). Very little rear overhang. Engine grilles on off-side behind the front axle. Centre or forward entrance. Most, but not all, carry the AEC triangle on the front panel. Single rear wheels.

General remarks:
A total of 319 Qs was built of which 233 went to London Transport. As a design it was well ahead of its time with most operators preferring to stick with conventional chassis layouts. By placing the engine on the offside behind the front axle AEC produced a design with the minimum engine intrusion. The transmission line to the rear axle was kept straight and operated through linkages via a fluid-flywheel and preselector gearbox to a differential on the offside. The complete bus weighed only six tons. Of the total number built only 23 were double-deckers produced from 1934-37 and the last of these was a six-wheel version supplied to the Green Line fleet. Fitted with hydraulic transmission and an electrically operated gearbox it was not very successful and soon saw demotion to Country Area bus duties, being withdrawn and sold in 1940. Four examples are in preservation, three of these being Birmingham B35C-bodied London Transport buses and the other a Willowbrook C37C-bodied coach supplied new to Silver Service of Matlock.

Below:
CGJ 188 is one of the preserved London Transport AEC Qs (Q83) carrying the distinctive Birmingham Railway Carriage & Wagon Company centre-entrance 35-seat bodywork and dating from 1935. The mid-underfloor engine is evidenced by the cooling grille and vents on the bodyside. Also to be noted is the rear axle set right back and allowing virtually no rear-overhang. Ian Allan Library

AEC Reliance and Monocoach

Built: 1953-79 (Reliance), 1953-57 (Monocoach)

Bodywork types: Single-deck bus and coach

Engines: AEC – 6-cyl 6.75 and 7.58 oil (Reliance from 1953)
6-cyl 9.6 oil (Reliance from 1961)
6-cyl 8.14 and 11.3 oil (Reliance from 1966)
6-cyl 12.47 (replaced other options in the Reliance from 1973)
6-cyl 7.58 oil (Monocoach)

Wheelbase lengths: 16ft 4in (Reliance from 1953 and Monocoach)
18ft 7in (Reliance from 1961)
20ft 0in (Reliance from 1968)

Typical bodywork makes:
Alexander, Duple, Harrington, Marshall, Park Royal, Plaxton, Roe, Weymann, Willowbrook

Recognition features:
Full-fronted, forward control designs, with bodywork mostly of manufacturer's standard designs. AEC triangle usually displayed on the front panel.

General remarks:
Designed as a successor to the Regal Mk IV, production of the Reliance ran alongside its predecessor for three years. The 16ft 4in wheelbase version was designed for 30ft 0in × 8ft 0in and, as the Construction and Use Regulations were altered, the wheelbase and chassis length were stretched to suit. The 18ft 7in wheelbase allowed for 36ft 0in bodywork and the 20ft 0in version for a body 39ft 4in, or 12 metres long. AEC was the first builder to offer a chassis suitable for 36ft 0in bodywork in 1961. Initially air or triple

---

Below:
An example of the 16ft 4in-wheelbase AEC Reliance is 9629 WU. 30ft 0in × 8ft 0in bodywork by Roe is fitted and the vehicle dates from 1960 when it was new to Felix Motors of Hatfield, South Yorkshire. It was operated by South Yorkshire PTE before coming into preservation.

servo vacuum brakes were offered with a five-speed synchromesh gearbox but in 1955 a four-speed direct air-operated epicyclic gearbox was offered. In 1956 the 6.75-litre engine was withdrawn. In 1961 a choice of four, five or six-speed synchromesh gearbox became available and in 1967 the alternatives became a six-speed synchromesh or epicyclic gearbox.

The Monocoach was a medium-weight underfloor-engined integrally constructed vehicle built in conjunction with Park Royal Vehicles. The only options offered were in transmissions — either a fluid-flywheel and four-speed epicyclic gearbox or friction clutch and five-speed synchromesh box. Air brakes were standard with vacuum servo as an option. Only 359 Monocoaches were built and production ended in 1957.

Top:
AMX 8A is a 1963 18ft 7in wheelbase AEC Reliance fitted with Harrington Cavalier coachwork. Bodywork is 36ft 0in × 8ft 0in. Now preserved it was new to Valliant of Bushey and then subsequently operated with Premier, Cambridge.

Below:
Similar but different AEC Monocoaches — GWG 472 has bodywork by Alexander whilst WRA 12 has the more normal Park Royal construction, both being B45F. They were side by side at the 1985 AEC Rally. M. Fowler

AEC Routemaster

Built: 1954-68

Bodywork type: Double-deck bus

Engines: AEC – 6-cyl 9.6 AV590 type oil
6-cyl 11.3 AV690 type oil
Leyland – 6-cyl 9.8 O600 type oil

Wheelbase lengths: 16ft 10in (RM, RMA, RMC, RMF)
19ft 2in (RML, RCL)

Bodywork makes:
All Park Royal vehicles (except prototypes RM3 [Weymann] and CRL4 — later RMC4 — [Eastern Coach Works])

Variants:
RM – 27ft 6in long open rear platform bus
RMA – 27ft 6in long forward entrance bus for British European Airways
RMC – 27ft 6in long Green Line coach version with rear platform doors
RMF – 30ft 0in long forward entrance bus for Northern General
RML – 30ft 0in long open rear platform bus
RCL – 30ft 0in long Green Line coach version with rear platform doors

Above:
RM11 (VLT 11) was an early production series Routemaster which, in this 1984 view in London's Victoria Street, retained its original style of air vent at upper-deck floor level and single panel upstairs front windows. In spite of being 25 years old by then it still looks a remarkably modern design.

Recognition features:
Low-set mesh radiator screen with chrome surround and AEC triangle at top of dividing strip. Bonnet curves down to the front with mudguards incorporated into the design. Bus versions have single-headlamps set into the front panels whilst the Green Line coach and BEA versions have twin headlamps with the innermost set slightly higher. Air intake situated over the cab, London Transport design bodywork. The long wheelbase version has a 2ft 4in section added at the halfway point, easily recognised by the small window on both decks.

General remarks:
Conceived as a London trolleybus replacement vehicle and first seen at the 1954 Commercial Motor Show, four prototypes were constructed for evaluation purposes. Two of these had AEC engines and Park Royal bodywork, the other two had Leyland O.600 engines, one being bodied by Weymann and the other by ECW to the standard London Transport design. Production commenced in 1958 and proceeded until 1968 by

---

Below:
The fourth preproduction Routemaster was RMC4 (SLT 59), originally CRL4, the prototype for later batches used on Green Line coach services. Built in 1957 it carries bodywork by Eastern Coach Works to full London Transport specification. After extensive use it has been preserved by London Country Bus Services and this view shows it at Hertford garage in 1980.

Right:
The first batch of lengthened Routemasters, recognisable by the short side windows on both decks, numbered 24 vehicles built in 1961. These were 72-seaters as opposed to the 64 seats of the standard design. The useful extra capacity led to a further 543 RMLs of which the first 42 were Green Line coach versions (RCL). RML900 (WLT 900) of the first batch, seen here in Uxbridge High Street, has since gone north of the Border to Clydeside Scottish and is now a regular exhibit at bus rallies.

Below right:
One of the former Northern General RMFs came to London in 1982 to work sightseeing tours with Obsolete Fleet. RMF2794 (FPT 588C) of 1965 was originally NGT3102 and carries a Park Royal 72-seat body with front entrance and platform doors. The small side windows used on the longer chassis are also evident in this view at Euston, as is the offside rear emergency door. By this time it had been fitted with standard London destination indicators.

which time 2,875 had been built. Of these 2,120 were standard RM and 524 were long wheelbase RML buses for London Transport, 68 were standard wheelbase and 43 long wheelbase Green Line versions. British European Airways took 65 forward entrance RMA buses whilst Northern General received 50 RMF buses. Transmission on all models is via fluid-flywheel and four-speed air-operated epicyclic gearbox, usually with fully-automatic control, to a fully floating rear axle. Whilst many are still in regular service increasing numbers are being preserved and a number have been exported.

AEC Bridgemaster and Renown

Built: 1956-63 (Bridgemaster), 1962-67 (Renown)

Bodywork type: Double-deck bus

Engines: AEC – 6-cyl 7.685 AV470 type oil (prototype Bridgemaster)
6-cyl 9.6 AV590 type oil (Bridgemaster and Renown)

Wheelbase lengths: 16ft 6in (Bridgemaster)
18ft 3½in (Renown)
18ft 4in (Bridgemaster from 1960)
18ft 10in (Bridgemaster from 1957)

Typical bodywork makes:
Crossley, East Lancs, Park Royal, Willowbrook

Variants: Bridgemaster MB3RA – prototypes
Bridgemaster B3RA – rear entrance production vehicles
Bridgemaster 3B3RA – forward entrance production vehicles
Renown 3B2RA – monocontrol gearbox
Renown 3B3RA – synchromesh gearbox

Recognition features:
Wide bonnets similar to that of the Regent V, inset mesh radiator screen with chrome surround and AEC triangle at top of central dividing strip. Single headlamps and nearside fog lamp set into front panels. Shallower windscreen and bodyside windows than on the Regent V. The Bridgemaster cab is noticeably higher than the lower saloon windows.

General remarks:
At the same time as the Routemaster was evolving AEC gave consideration to the production of low-height design double-deckers. This resulted in the Bridgemaster displayed at the 1956 Commercial Motor Show. This carried a Crossley badge as that company had undertaken the development work. The Bridgemaster was designed as a chassisless vehicle with subframes front and rear and coil springs. Braking was hydraulic at the front and air at the rear, transmission was via a four-speed synchromesh gearbox. In 1956 the AV470 engine was replaced by the larger AV590 and in 1957 a longer wheelbase enabled 30ft 0in bodywork to be used. The later 18ft 7in wheelbase allowed for 30ft 0in forward entrance bodywork.

The Renown was designed as a Bridgemaster replacement, having a conventional chassis and low floor, dropped-centre rear axle and rear air suspension. The AV590 engine was fitted as standard and either four-speed monocontrol epicyclic or synchromesh gearboxes were optional. Air pressure braking was standard. Some 180 Bridgemasters and 252 Renowns were built and several examples of both have been saved for preservation.

Above right:
The AEC Bridgemaster and Renown were superficially very similar vehicles, both having lowbridge bodies and Regent V style radiator grilles. East Yorkshire was the largest operator of the Bridgemaster with 50 vehicles bought in 1960-63. Bridgemaster features include a Regent V type grille and bonnet, vertical offside lower front panel, high driver's step and small offside lower-deck front window. The EYMS examples such as 742 (3742 RH) also had sloping upper-deck side windows to allow their passage under Beverley Bar.

Right:
The Renown was fitted with an external sliding driver's door, lower step and five equal width offside lower-deck side windows. The offside mudguard projected slightly from the front panel. Driving cab windows are larger than on the Bridgemaster. East Yorkshire 782 (CKH 782C) was one of 34 bought between 1964 and 1966.

HULL
14
742
EAST YORKSHIRE
3742 RH

HULL
Driffield
12
Matador for Clarksons holidays
EAST YORKSHIRE
782
CKH 782C

Albion — the 1920s — 24, 26 and Viking 28 Series

Built: 1923-32

Bodywork types: Single-deck bus and coach

Engines: Albion – 4-cyl 24hp and 29.8hp petrol (24, 26 and 28 series)
6-cyl 45hp petrol (28 series)

Wheelbase lengths: 10ft 9in (PA24)
11ft 3in (PB24 and PC24)
12ft 9in (PE24)
14ft 0in (PF24 and PH24)
16ft 3in (Viking PM28, PMA28, PMB28, PFB26, PKA26, PKB26, PNA26)

Typical bodywork makes:
Mumford, Roberts, Roe, Strachans, Vickers plus many local coachbuilders

Recognition features:
Normal control chassis except for the PM28, PMA28 and PMB28. Large squarish radiator incorporating the Albion rising sun motif on the toptank. Headlamps and dumb-irons well forward of the radiator front. Front mudguards separate from the bodywork. The 26 series has a slightly narrower radiator with decorative mouldings.

General remarks:
The Albion Motor Car Company Ltd introduced its first true bus chassis in 1923 after a long period in which many of its goods chassis had received bus and charabanc bodies. This was the 24 series derived from the 25cwt and 30cwt goods chassis. It was produced in a variety of wheelbases suitable for 12/21 seat bodywork in a normal control layout. In 1925 the more powerful 26 range was introduced, again in a variety of wheelbase lengths. In the following year a more powerful engine was offered in the 24 series. Like the 24 the 26 was of normal control design but in 1927 the first forward control model, the PM28, was introduced. Most were powered by the 90bhp six-cylinder engine but the four-cylinder unit was an option. In the same year the normal control PNA26 appeared. The PFB26 was fitted with front-wheel brakes, the PMA28 had a four-cylinder engine and the PMB28 the six-cylinder unit, the PKA26 and the PKB26 being the normal control equivalents. Production of the 24 series ended in 1929, the 26 and 28 series were phased out in 1931-32. Examples of the PB24, PH24, PK26, PM28 and PMA28 have been preserved.

Above right:
ES 5150 is an Albion PB24 dating from 1926 which, fitted with a 13-seat body, was used as a Scottish hotel bus. Now in the care of Western Scottish it was photographed being rallied at a Strathclyde open day. The Albion motif is clearly visible on the radiator.
Stewart J. Brown

Right:
This Albion PM28 was operated by Leicester & District and is typical of its type with the radiator set well back, triangular-topped bonnet and very high steering wheel. RY 4520 was photographed at Skegness in July 1929. G. H. F. Atkins

PRIDE O'TANNOCHBR
1
047 SD

PRIVATE
LEICESTER & DISTRICT
GREEN BUS CO

Albion — 1930s — Valkyrie, Victor, Valiant, Venturer and Valorous

Built: 1930-39

Bodywork types: Double-deck bus, single-deck bus and coach

Engines:

Albion –	4-cyl	3.15	petrol (Victor PH49)
	4-cyl	3.62	petrol (Victor PH49)
	4-cyl	3.89	petrol (Victor PH115)
	4-cyl	5.0	petrol (Valkyrie PX65 and PW65)
	4-cyl	6.1	petrol (Valkyrie PW67, PW69, PW141 and PR145)
	6-cyl	6.03	petrol (Valiant PV70)
	6-cyl	6.1	petrol (Valkyrie CX9)
	6-cyl	6.85	petrol (Venturer 80)
	6-cyl	7.8	petrol (Valkyrie PV141, PW145, Valiant PV71, Venturer 81 and Valorous 85)
	6-cyl	9.1	petrol (Valkyrie CX13)
	4-cyl	6.6	oil (Valkyrie CX9)
	6-cyl	9.1	oil (Valkyrie CX13 and Venturer CX19)
Gardner –	4-cyl	5.6	4LW oil (Valkyrie CX9)
	5-cyl	7.0	5LW oil (Valiant PV71, Valkyrie CX11 and Venturer 80)
	6-cyl	8.4	6LW oil (Venturer CX19)

Wheelbase lengths: 14ft 6in (Victor — all models except PK114)
16ft 3in (Valkyrie PX65, Venturer — all models, Valorous)
17ft 2in (Victor PK114)
17ft 4in (Valiant PV70 and PV71)
17ft 8in (Valkyrie — all models except PX65)

Typical bodywork makes:
Alexander, Brush, Cowieson, Duple, English Electric, Northern Counties, Roberts, Strachans Successors Ltd, Willowbrook

Recognition features:
All single-deck except the Venturer and Valorous. Earlier examples have old style Albion radiator design with wide framing. This was developed into a slimmer, taller design with a reshaped Albion badge on the toptank. Most have the Albion and vehicle name on the radiator mesh.

General remarks:
Albion introduced the Valkyrie in 1930 as a forward control, 16ft 3in-wheelbase chassis for 32-seat bodywork. Designated PX65 it was soon followed by the 17ft 8in wheelbase PW65 for 36-seat bodywork. In 1933 the PW67 replaced the 65 series and in 1934 the 39-seat PW69 was introduced. The PW141 using the 6.1-litre engine and the PV141 using 7.8-litre unit followed in 1935. In 1936 the PR145 and PW145 three-axle versions made a brief appearance.

The normal control PH49 Victor was introduced in 1930 to replace the PH24 and in 1933 was joined by the larger PHA49 and PHC49. In 1934 they were replaced by the PHB49 whilst the 26-seat PH111 was added. Also in that year the new 3.89-litre engine produced the PH115, a forward control chassis with the same wheelbase as the PH49. A longer wheelbase forward control PK115 was also introduced and the normal control PH114 and PK114. Production ended in 1939.

The Valiant PV70 made its appearance in 1931. Designed for 36-seat bodywork it used the 6.03-litre engine but in 1935 became the PV71 using the 7.8-litre or Gardner 5LW engine. Production ended in the following year.

In 1932 Albion launched its first double-deck chassis, the Venturer 80, with a 16ft 3in wheelbase and either Albion petrol or Gardner diesel engine. In the following year a larger 6.98-litre Albion petrol engine was offered. The Venturer 81 of 1935 brought the seating capacity up from 51 to 55 and used the Gardner 6LW engine as standard — the Albion 7.8-litre petrol unit being an option. A three-axle version was built in 1932 and

Top:
The 1930 Albion Valkyrie was visually akin to the earlier PM28 series as shown in this view of WG 1448 at Brighton. An Alexander B34F-bodied PW65 was new to the Alexander group of companies and is now preserved. L. H. Page

Above:
AAA 756 is an Albion Victor PK114 with Abbott C20C body, new to King Alfred Motor services, Winchester, in 1935. The revised radiator is evident along with the Albion Victor name on the mesh. M. A. Penn

---

named the Valorous 85. This did not go into production and the only example built was sold to Young's of Paisley after being used as a demonstrator.

In 1937 the Valkyrie and Venturer were superseded by new models in the CX range which had engines and gearboxes installed as a single unit with a range of engine options. Production ceased at the beginning of World War 2 but resumed in 1945 with oil engines as standard. Production of the Valkyrie ended in 1950 and the Venturer — which became the CX37 in 1947 — in 1951.

All models used a single-plate clutch and either a sliding or constant mesh gearbox. Examples of the Valkyrie PW65, PW141 and CX13, Victor PH115 and PK114, and Valiant PV70 have been preserved.

Albion — Postwar Front-Engined Types

Built: 1945-78

Bodywork types: Double-deck bus, single-deck bus and coach

Engines:

Albion –	4-cyl	6.1	petrol (Valkyrie CX9)
	6-cyl	4.26	petrol (Victor FT3AB)
	6-cyl	9.1	petrol (Valkyrie CX13)
	4-cyl	4.88	oil (Victor FT39)
	4-cyl	5.48	oil (Victor FT39K, VT15)
	4-cyl	6.6	oil (Valkyrie CX9)
	6-cyl	9.1	oil (Valkyrie CX13, Venturer CX19)
	6-cyl	9.9	oil (Venturer CX37, Valiant CX39, Viking CX41/HD61)
Gardner –	4-cyl	5.6	4LW oil (Valkyrie CX9)
	5-cyl	7.0	5LW oil (Valkyrie CX11)
	6-cyl	8.4	5LW oil (Valkyrie CX13, Venturer CX19)
Leyland –	6-cyl	5.76	O.350 oil (Victor VT17)
	6-cyl	6.17	0370 oil (Victor VT17B, VT21L, Viking VK41)
	6-cyl	6.54	O.400 oil (Viking VK43, VK49, VK55, VK57, VK67)
	6-cyl	9.8	O.600 oil (Lowlander)

Wheelbase lengths:

12ft 5in	(Viking EVK55CY)
15ft 6in	(Victor VT15N, VT17N, VT17B)
16ft 0in	(Victor FT39AN)
16ft 0⅝in	(Venturer CX)
16ft 1in	(Viking VK41)
16ft 2in	(Viking VK43, VK49, VK57, VK67)
16ft 11½in	(Victor FT39AL, FT39K)
17ft 3in	(Victor VT15L, VT15N, VT17B)
17ft 7in	(Valkyrie CX9, CX13, Valiant CX39)
17ft 8½in	(Victor VT21L)
18ft 3in	(Viking CX41)
18ft 6in	(Lowlander)
19ft 0in	(Viking HD61)
20ft 0in	(Viking EV55CUL)
20ft 6in	(Viking CX41)
21ft 0in	(Viking HD61)

Below left:
Production of the Albion Venturer CX19, first introduced in 1938, was resumed in 1945. This 1949 CX19 with Metro-Cammell-Weymann H30/26R bodywork, spent its whole working life as Cheltenham and District 72 and is preserved locally. The vehicle carries the standard postwar radiator with the 'Albion Oil Engine' badge. HDG 448 was photographed at the 1974 Bristol Rally. C. L. Caddy

Above:
The postwar Victor FT was very popular with the Guernsey Railway Co whose No 20 (1787) is seen here after preservation. The radiator is modified with a distinct 'rising-sun' motif and bodywork is B36F by Reading of Portsmouth. The chassis is an FT39AN.
J. G. Lidstone

Typical bodywork makes:
Alexander, Brockhouse, Duple, Guy, Harrington, Heaver, Metro-Cammell, Mulliner, Northern Counties, Pennine, Reading, Roberts, Roe, Strachans, Thurgood, Whitson, Weymann

Recognition features:
All forward control models with parallel-sided radiators having the Albion rising sun motif at the top. Many also carry the model name on the radiator mesh. Guy bodywork has distinctive outswept skirt panels. The Victor VT and Viking VK models have concealed radiators in the bodymakers' usual styles but display the Albion motif on the front panel. The Lowlander has the St Helens front concealed radiator with either the Albion or Leyland name across the centre of the grille.

General remarks:
Production of Albion models resumed in 1945 with the Valkyrie CX9 and CX13 and the Venturer CX19. All had oil engines as standard. In 1947 the Venturer was developed as the CX37 by using the larger 9.9-litre engine. Production of these ended in 1950 (Valkyrie) and 1951 (Venturer).

The Victor FT was introduced in 1947 as a new straight-framed lightweight chassis. The first model was the FT3AB using the 4.26-litre engine and was joined in 1948 by the FT39 series using a 4.88-litre oil engine. A longer version introduced in 1951 was designated FT39AN and a yet longer variant in 1954 the FT39AL. In 1956 the larger 5.48-litre engine was installed to make it the FT39KAN. Production ended in 1959.

In 1948 the Valiant CX39 replaced the Valkyrie CX13 and used the same engine as the Venturer. Valiants were mostly fitted with coach bodywork and production only ran until 1951. 1948 also saw the debut of the Viking CX41 intended mainly for export and having

Above:
The majority of Albion Valiant CX39Ns were fitted with coach bodywork as on this 1950 example, FVA 854, which was new to Hutchinson, Overtown. The Duple body is the standard design of its period with raked windscreen, curving window-line and with no canopy over the engine. The flared skirt and swept mouldings should also be noted.
M. D. Shaw

Below:
The Albion Lowlander looked superficially similar to the St Helens front Leyland Titan PD3 and certainly the radiator and bonnet assembly was virtually identical. Most Scottish Lowlanders carried the Albion name whilst English ones were Leyland badged. Differences were evident in the single-piece windscreen and sliding passenger door of the Lowlander. South Notts 82 (now preserved) and 89 are Lowlander LR3s with Northern Counties bodywork seen here at the depot in April 1983.

a straight-frame and four-speed gearbox. In 1950 it was upgraded to the HD61 by the installation of a five-speed gearbox.

When production of the Victor FT ended in 1959 it was replaced by the VT series, the VT39 using an Albion and the VT17 a Leyland engine. With chassis modifications 'A' suffixes were added. In 1962 the Leyland-engined model became the VT17B. A longer VT21L was introduced in that year but lost out to the dominance of Bedford and production ended after a few months. The following year the thoroughly modern Viking VK41 came on to the market. This had a set-back front axle, air-hydraulic braking and a Leyland engine mounted on the front of the main frame. The VK55 with a larger engine and full airbraking followed and in 1965 the VK43 which was redesigned with a rear-mounted engine. This was developed as the VK49 (pneumocyclic gearbox), the VK57 (Turner gearbox) and VK67 (full airbraking). Production of the Viking ended in 1978. Albion's final double-deck model was the Lowlander of 1961, introduced mainly to meet Scottish Bus Group needs. It had a drop-centre rear axle and came in four variants — the LR1 (pneumocyclic gearbox), LR3 (synchromesh gearbox), LR5 and LR7 (same as LR1 and LR3 but with air-suspension on the rear axle). In England it was usually marketed as a Leyland vehicle. Production ceased in 1966.

Preserved Albions include the Valkyrie CX13, Victor FT3AB, FT39N, FT39AN, FT21N and VT21L, Valiant CX39N, Viking VK43L and Lowlander LR1 and LR3.

Albion Underfloor-Engined Chassis

Built: 1951-60

Bodywork types: Single-deck bus and coach

Engines: Albion – 4-cyl 3.83 oil (Nimbus MR9
4-cyl 4.1 oil (Nimbus NS3N, NS3AN, NS3L)
8-cyl 9.76 oil (KP71NW)
Leyland – 6-cyl 5.76 O.350 oil (Aberdonian MR11L)

Wheelbase lengths: 11ft 10in (Nimbus)
16ft 4in (KP71NW and Aberdonian)

Typical bodywork makes:
Alexander, Harrington, Plaxton, Seddon, Strachans, Willowbrook

---

Below:
BSD 470 is one of only two Albion KP71NWs to be built and has survived to be preserved. This photograph in June 1963 shows it at work in its new guise. It was bought back from Western SMT by Albion Motors for preservation and carries the last body to be built by Scottish Aviation. H. Hay

Recognition features:
Forward control, full-fronted single-deckers. The Albion motif appears on the front panel.

General remarks:
Albion introduced its first underfloor-engined design in 1951 with the experimental KP71NW which used an eight-cylinder engine comprising two horizontally-opposed four-cylinder blocks. Only two were ever built, both being bodied by Scottish Aviation of Prestwick. The next design was the lightweight Nimbus which made its appearance in November 1955. Designated MR9 it used the 3.83-litre engine but this was replaced in 1958 by the NS3N and NS3L which were 23ft 3in and 24ft 0in long respectively and used the new 4.1-litre engine. In 1961 further modification in the form of a five-speed gearbox produced the NS3AN. The Nimbus sold in relatively small numbers to rural operators and was withdrawn in 1963. In the summer of 1957 Albion introduced the Aberdonian MR11L which was a more conventional chassis. On an average length wheelbase it used the Leyland O.350 engine together with a five-speed constant mesh gearbox and vacuum hydraulic brakes. Many operators purchased one-off examples and a few were exported, production ceased in 1961. Examples of all except the Nimbus NS3AN are now preserved.

Below:
Devon General was an Albion Nimbus operator and its Willowbrook B31F-bodied XTA 839 (Fleet No SN839) is now preserved in its operating area. The Willowbrook bodywork has similarities to the ECW standard design of the time. Note the extremely short 11ft 10in wheelbase and the 8ft 0in body width making the bus look very short. XTA 839 is an MR9N of 1958, the year in which the model was superseded by the NS3 series.
Roger Bailey

Above right:
It could be a Tiger Cub or a Regal but the Albion badge gives it away as an Aberdonian MR11L — 674 (VKH 674) of East Yorkshire — at Showbus 1984 collecting one of the awards. Dating from 1958 it was, like so many operator's purchases, a one-off, and ran in service until 1969 when it was converted into a left luggage office at Scarborough until bought for preservation in 1975. Bodywork is B39F by Park Royal.

Atkinson Alpha

Built: 1950 (PM series), 1953 (PL series), 1955 (PD prototypes). PM and PL series production ended in 1964

Bodywork type: Double and single-deck bus

Below:
A pair of Atkinsons in the SHMD fleet at its Dukinfield depot in 1968. On the left UMA 370 is a PD746 of 1955 with Northern Counties H59CD bodywork and next to it XLG 478 is an Alpha PL746H of 1956 with B34C+27 bodywork by Metro-Cammell. UMA 370 and a similar PL746H — XLG 477 — have been preserved. The Atkinson symbol appears prominently on the double-decker's radiator. A. Moyes

Engines: Gardner –	4-cyl	5.6	4HLW type oil (PL744H)
	5-cyl	7.0	5HLW type oil (PL745H and PM745H)
	6-cyl	8.4	6HLW type oil (PL746H and PM746H)
	6-cyl	8.4	6LW type oil (PD746)

Wheelbase lengths: 16ft 6in (all models)

Typical bodywork makes:
Marshall, Northern Counties, Weymann

Recognition features:
The Atkinson symbol of an encircled capital A on the radiator or front panel. Single-deck models all had full-front bodywork, the double-deck prototypes having half-cabs and 'tin-front' concealed radiators with Atkinson symbol thereon. The preserved Alpha PL and the double-deck PD746 both have centre-doorway bodywork.

General remarks:
Atkinson & Co began in 1916 by selling Sentinel steam wagons, later developing its own models. In 1932 the company began producing its own internal combustion-engined trucks but it was not until 1950 that the first PSV chassis appeared. The Alpha PM was a medium-weight single-deck chassis with an underfloor-mounted horizontal engine and a David Brown five-speed constant mesh gearbox. The PM746H using the Gardner 6HLW engine was followed in 1952 by the PM745H using the 5HLW unit. In 1953 the Alpha PL was introduced as a lightweight underfloor-engined single-deck chassis for bodywork of up to 30ft 0in long. A choice of three engines, the Gardner 4HLW, 5HLW and 6HLW, was available, designations being PL744H, PL745H and PL746H. The same gearbox and running units were employed as on the Alpha PM series. In 1955 Atkinson attempted to break into the double-deck market with the PD746 using the same wheelbase as the single-deck chassis. Originally offered with a four-speed constant mesh gearbox, in 1957 a four-speed epicyclic unit became available. However, in spite of being on the market until 1963, only two prototype chassis were ever built, one of which is now preserved. Fortunately both types of single-decker are represented by preserved examples.

Austin

Built: 1923-55

Bodywork type: Single-deck

Engines: Austin –	4-cyl	22.4hp	or 6-cyl 23.4hp petrol (Twenty)
	6-cyl	3.5	petrol (CXB)
	6-cyl	4.0	petrol (CXB, K8, CXD, K2SL, K4SL)
Perkins –	6-cyl	4.7	P6 oil (CXD)

Wheelbase lengths: 10ft 10in and 11ft 4in (Twenty)
11ft 2in (K2 and K2SL)
13ft 1½in (K4 and K4SL)
7ft 9in (K8)
15ft 0in (CXB and CXD)

Typical bodywork makes:
Jefferies, Mann Egerton, Metalcraft, Plaxton, Reading, Whitson

Recognition features:
On normal control models the bonnet is superficially similar to that of the famous Bedford OB, although somewhat smaller, namely a rounded top and radiator grille, forward sloping and slightly upswept mudguards, headlamps mounted on the bonnet sides. On forward control, full-fronted, models the radiator grille features central, vertical slotting flanked by horizontal slots in a basically oval frame. The standard design as used on the K8 has a distinctly rounded top on curving front-panel and headlamps incorporated into the bodywork. The Austin name appears on the top of the grille on all models.

General remarks:

The Austin Motor Co began production in 1910 and an interesting survivor from that period is a 1911 Estate Saloon registered C 78 which is privately preserved. Although not strictly a PSV neither is it a car in the normal sense. For both car and light commercial applications the Twenty was introduced in 1923, soon proving a popular chassis it remained in production until 1940. The first designated PSV chassis arrived in 1939 and was a 26-seat 1½ to 3 ton vehicle. At first given the chassis type K2 and K4 it

---

Below:
The CXB of 1947 was Austin's 'answer' to the Bedford OB. This example demonstrates the somewhat similar appearance of the CXB but emphasises the much smaller bonnet and radiator. Bodywork is by Plaxtons of Scarborough for Jones of Wolverhampton. British Leyland

Bottom:
The K2SL of 1948 found many applications in rural bus work such as DST 358 seen in the Scottish Highlands at Strathcarron. Dating from 1949 it has bodywork by Federated Industries for 17 passengers. D. A. Lawrence/A. Moyes collection

Above:
The last Austin bus was the CXD. It was introduced in 1948 and found use mainly with coach bodywork as typified by this example belonging to R. Dickson of Dundee. Its Plaxton bodywork emphasises the chrome finishing so popular in 1952 when it was new. Plaxtons

---

was modified to the K2SL and K4SL in 1948. In its later forms it was primarily intended for the goods market but many received passenger bodies during the immediate postwar vehicle shortages. The CXB of 1947 was Austin's 'answer' to the Bedford OB whilst at the same time the small K8, designed primarily as a light goods chassis, found numerous passenger applications. The K8 was Austin's first forward control design, the CXB being normal control and looking very much like the Bedford OB which it challenged. The CXD was also forward control and was to remain in production until 1955 as the CXD series II with Perkins engine.

Beadle

Built: 1946-55

Bodywork type: Single-deck

Engines Bedford – 6-cyl 27.34hp petrol, and others according to basis of vehicle

Wheelbase lengths: 17ft 6in (AEC, Bedford and Leyland derivatives)
17ft 11in (Leyland derivatives)

Typical bodywork make:
Beadle

Recognition features:
All examples built to full-fronted design, the Bedford derivatives having two vertical grilles set side by side, both having vertical slotting with oval tops and bottoms. Most Leyland derivatives have front panels and radiator grilles similar to the full-fronted Leyland design of the period, namely two vertical grilles with horizontally curving bars across and surmounted with curved bars aligning with the lower windscreen. All have four-piece windscreens.

Above:
The early Beadle buses based on Bedford running units had very austere bodywork as seen in this Eastern National example. Note the style of radiator grille used on all Beadle-Bedfords, the access door to the engine compartment and the roof-line — very similar to Beadle's bodywork fitted to the Bedford OB. Ian Allan Library

Below:
When Beadle commenced using Leyland running units for coaches in 1951 the body design was considerably updated to this attractive style. GFN 273, now preserved, is a Beadle-Leyland CF35F of East Kent which used components from a prewar Leyland Titan TD5. C. L. Caddy

General remarks:

The Dartford firm of Beadle Coachbuilders began, in 1946, to produce integrally constructed single-deck vehicles using Bedford mechanical components and running units. In 1950 the use of components from AEC Regents was begun but discontinued after only a short period. In 1951 a programme using components from prewar Leyland vehicles — Tigers and Titans — commenced. As with the Bedford derivatives these were medium-weight vehicles. Alongside its medium-weight production the company had begun, in 1948, to produce a lightweight version using components from prewar Leyland Cubs and these, together with other Leyland derivatives, continued in production until 1955. The Beadle could be found in many rural operators' fleets and examples of all three types are preserved.

Above:
Lincolnshire Road Car 699 (EFU 843) was a typical example of Beadle's design based on the Leyland Cub. New in August 1948 it was operated until 1957 whereupon it found new use with a contractor in St Albans. It retains its Cub-radiator complete with Leyland name. R. H. G. Simpson

Bean

Built: 1925-32
Bodywork type: Single-deck
Engine: Bean – 4-cyl 13.9hp petrol
Wheelbase lengths: 11ft 1¼in (increased to 11ft 1½in in 1928)
Typical bodywork makes:
Birch Bros Ltd, Willowbrook
Recognition features:
Normal control chassis designed for 16-seat bodywork. Angular, vertical-sided bonnet with vertical louvres, oval top to bonnet, slots in side frame of radiator which has the Bean name on an oval plate on the top tank. A prominent crash-bar across the radiator mid-point is supported by side irons set at an angle. Low-set headlamps mounted on the frame beside the bonnet.
General remarks:
A. Harper Sons & Bean Ltd of Tipton, Staffordshire, entered the vehicle manufacturing world in 1919 and in 1924 produced a 1½-ton commercial chassis which was to prove popular with van, truck and bus operators. In 1927, after being reformed as Bean Cars Ltd, the company introduced a sturdier 1½-ton derivative. Financial difficulties caused the bankruptcy of the company in 1932.

Above:
There was really only one design of Bean and this is it! Registered 2377 this was Guernsey Railway Co No 19 with typically angular 16-seat bodywork and a fold-back roof.

Bedford W Series

Built: 1931-33 (WHB), 1931-35 (WLB), 1931-39 (WLG), 1934-39 (WTL), 1935-39 (WTB)
Bodywork type: Single-deck bus and coach
Engines: Bedford – 6-cyl 3.18 petrol (WHB, WLB, WTL, WTB)
6-cyl 3.519 petrol (WLG, WTL, WTB)
Wheelbase lengths: 10ft 11in (WHB)
11ft 11in (WLG from 1935)
13ft 1in (WLG, WLB, WTL)
13ft 11in (WTB)

Typical bodywork makes:
Burlingham, Duple, Economy, Heaver, Rainforth, Weymann, Willmott

Recognition features:
Narrow bonnet tapering down to a thin radiator having straight sides and rounded top and bottom, the Bedford name usually appearing on the mesh. Oval cutaways on inside edges of mudguards of WHB, WLB and WLG. Flimsy looking bumpers, flat windscreens and angled quarter-lights. The WTB introduced a rounded radiator grille with horizontal cuts and vertical divider, and horizontal ventilation grilles on the bonnet sides. Mudguards were larger but bumpers still flimsy. Headlamps alongside the radiator on all models. All normal control.

General remarks:
The WHB, introduced in 1931, was the first of two passenger chassis marketed together under the Bedford name. Designed for 14-seat bodywork it had a four-speed sliding-mesh gearbox and mechanical brakes. The other chassis was the WLG designed for 20-seat bodywork. Using the same engine and gearbox it had cable-operated brakes. In 1935 the wheelbase was shortened and the larger 27.3hp engine installed. However, most WLGs were sold for goods application. Later in 1931 the 20-seat WLB was introduced using the same engine, gearbox and mechanical brakes as the WHB. In 1934 the 3-ton goods model, the WTL, appeared and was soon modified to take passenger bodywork of 20-seats; it differed from its forerunners in having servo-assisted mechanical brakes. In 1937 the 3.519-litre engine was fitted. The final model in the series was the more modern-looking WTB of 1935 which, again in 1937, was increased in power by fitting the 3.519-litre engine. In other respects it was mechanically similar to the WTL. Total production was in excess of 5,000 of which one WHB, one WLG, seven WLBs, two WTLs and 13 WTBs have been preserved.

Below left:
TM 9347 is a good example of the diminutive Bedford WHB of 1931. It has flat, sloping windscreens, tapering bonnet and a radiator set well back. Only the dress of its occupants give the picture away as being a recent one — the bus is happily preserved in North London. It has Waveney bodywork. Vauxhall Motors Ltd

Above:
The WLB was a 20-seater and CMG 30 has coach-seated Duple bodywork. It was one of the last to be built in 1935, was new to Garner of Ealing but is now preserved by E. B. Taylor of Tintinhull in Somerset.

Below:
Much more modern in appearance was the Bedford WTB. JT 8077, which was new to South Dorset Coaches in 1937, has Duple C25F bodywork and now resides in the Forest of Dean. The Bedford emblem is carried on the radiator toptank and the name on the mesh. D. P. Buley

Bedford ML and O Series

Built: 1939/45-53 (ML and OLAZ), 1939/45-51 (OB), 1939-45 (OWB)

Bodywork type: Single-deck bus and coach

Engines: Bedford – 6-cyl 3.519 petrol
Perkins – 6-cyl 4.7 P6 type oil (OB option from 1945)

Wheelbase lengths: 11ft 11in (ML
13ft 1in (OLAZ)
14ft 6in (OB and OWB)

Typical bodywork makes:
Barnaby, Duple, Lee, Mulliner, Plaxton, Reall, Roe, SMT, Thurgood

Recognition features:
Low, rounded bonnet and radiator, both having horizontal ventilation louvres. Headlamps each side of bonnet, flimsy-looking bumpers, two-piece flat windscreens slightly vee-shaped. Rounded bodywork except wartime OWB which has the angular 'utility' style. all normal control designs.

General remarks:
The ML was introduced in 1939 as a goods chassis for 2/3-ton payloads; most were used for goods applications but a number were bodied as buses and coaches. A four-speed gearbox and fully floating rear axle together with Lockheed vacuum servo-assisted hydraulic brakes acting on all wheels were fitted as standard. In the same year the OLAZ appeared as one of 13 variants of the O series chassis. Designed mainly for goods application in the 3/4-ton range it was mechanically similar to the ML.

Also in 1939 the famous OB made its debut as a direct replacement for the WTB but production was soon halted by the outbreak of hostilities. Resuming in 1945 a remarkable total of 12,766 was built before production ended in 1951. Designed for 26/30-seat bodywork it was fitted with a four-speed gearbox using spur gears, a dry-plate clutch and fully floating rear axle. Braking was of the vacuum servo-assisted hydraulic type. Most were fitted with Duple Vista bodywork and almost 200 are now in preservation.

Bedford was chosen in 1939 as the wartime manufacturer of single-deck chassis and the wartime variant of the OB, the OWB, saw a production total of 3,398 of which 10 are now preserved. Mechanically it was identical to the OB but bodywork, to an almost identical design, was supplied by a variety of builders.

Top right:
The Bedford ML and OLAZ replaced the WTB and were direct antecedents of the classic OB. An OLAZ with Duple B20F bodywork is KGD 904 which operated with MacBrayne's (No 163) from new in 1952 and is now preserved by its Belfast owner. Ian Allan Library

Above right:
FSA 232 is an absolutely typical Bedford OB dating from 1950. It also has typical Duple Vista C26F bodywork of the type fitted to the majority of OBs built. It was first registered in Aberdeen but is seen here in semi-preservation in Banbury in 1987. Melanie Wise

Right:
GZ 783 is an absolutely typical Bedford OWB with utility 27-seat body by Scottish Motor Traction. It was built in 1942 for its Northern Ireland owner and still resides in the province. S. C. Morris

MACBRAYNES

ADKINS
MYSTERY
FSA 232

NORTHERN IRELAND TRANSPORT
GZ 783

The Duple Bella Vega body was in a completely different style to its predecessors and an example was carried by this Bedford SB5 belonging to Clue's Coaches of Menheniot, seen here on a Liskeard market day service in 1984. Note the Bedford badges, the Bella Vega name and the rear side-window lines.

Bella
Vega
529 PTT

Bedford SB Series

Built: 1950-86

Bodywork types: Single-deck bus and coach

Engines: Bedford – 6-cyl 4.927 petrol (SB, SBG, SB3)
6-cyl 4.927 oil (SB1)
6-cyl 5.42 oil (SB5)
6-cyl 6.095 oil (SB13)
Leyland – 6-cyl 5.76 O.350 type oil (SB8)
Perkins – 6-cyl 5.562 R6 type oil (SBO)

Wheelbase lengths: 17ft 2in
18ft 0in (from 1955)

Typical bodywork makes:
Duple, Marshall, Strachans, Thurgood, Willowbrook

Recognition features:
Earlier examples had a bulbous front cowl incorporated into the body builders own design, the Bedford name being displayed on a horizontal chrome bar at the top of the radiator grille. Later examples tended towards a one-piece wrap-round windscreen, flatter fronts and wider grilles. The Bedford insignia is usually prominent on the front panel. Many examples bodied by Strachans and Willowbrook were supplied for military use.

General remarks:
The SB, Bedford's first forward control design, was launched in 1950 as the replacement for the OB and, as a 33-seater, it was the company's largest design to date. It used a four-speed synchromesh gearbox and hypoid rear axle, braking being of vacuum servo hydraulic type. In 1953 the Perkins engine was offered, this version being designated SBO, whilst at the same time the SB was redesignated SBG. In 1955 production moved to the new Dunstable plant at which point the wheelbase was lengthened. In 1957 the availability of a Bedford diesel engine and the Leyland O.350 brought into being the SB1 and SB8. In 1959 the SBO was discontinued and in 1959 the SBG was reclassified SB3. The SB5 appeared in 1962 using the larger Bedford oil engine and at the same time the SB8 was fitted with hydraulically operated air brakes. In 1963 the SB8 was replaced by the SB13 using a still larger Bedford oil engine and in that year the remaining variants were offered with five-speed synchromesh gearbox. From 1968 the General Motors computer classification for the SB series became NJM but they were rarely referred to under that code. Examples of the SB, SBG, SBO, SB1, SB3, SB5 and SB8, totalling over 50, have been preserved.

Above right:
The Bedford SB3, launched in 1950 as the SB, redesignated SBG in 1953 and SB3 in 1959, was one of Bedford's longest-running designs. This example — WTX 87 — was built as recently as 1965 and fitted with a Duple Midland B42F body typical of the type used by many operators for schools contract work. Osmonds of Curry Rivel, Somerset, were operating it when seen in 1985 having bought it from Turburville, Hemel Hempstead in 1969.

Right:
A Duple Super Vega C41F body was carried by this 1962 Bedford SB5 owned by Wake's of Sparkford at whose Wincanton premises it was photographed in 1985. It had been bought by them from Byng, Portsmouth in 1963. The very rounded lines of the Duple bodywork are notable, as is the Bedford name and badge and the fairing over the rear wheelarch.

SCHOOL BUS
OSMONDS
WTX 87

YRV 301

Bedford V Series

Built: 1961-70

Bodywork types: Single-deck bus and coach

Engines:

Bedford –	6-cyl	3.52	petrol (VAS)
	6-cyl	4.927	petrol (VAM)
	6-cyl	4.927	oil (VAM, VAS)
Leyland –	6-cyl	6.54	O.400 type oil (VAL)

Wheelbase lengths: 13ft 8in (VAS)
16ft 1in (VAM)
17ft 8in (VAL)

Typical bodywork makes:
Duple, Plaxtons

Recognition features:
The VAS, like its SB predecessor, has the front axle set right forward with the doorway behind. The others have set-back front axles allowing a forward-placed doorway. Bedford insignia prominent on front panel. The VAM has two axles, the VAL three axles with twin-steering and small wheels. All are forward control designs.

General remarks:
The VAS was launched in 1961 with a choice of petrol or diesel engines and a chassis length suitable for 29-seat bodywork. The VAS pioneered the use of small wheels (16in diameter) allowing a lower floor height and improved access for passengers. The VAL which followed in 1962 used the same size wheels and had the same floor height but was aimed at the 55-seat coach market using 36ft 0in-long bodywork. To make this possible it was designed as a three-axle vehicle using four-wheel steering on twin front axles. Fitted with the Leyland O.400 engine it had a five-speed synchromesh overdrive gearbox, the engine being mounted on the front overhang. It proved popular with operators, passengers and enthusiasts alike. The more conventional VAM followed in 1965 and was a 10m (32ft 10in) vehicle with the engine mounted similarly to that of the VAL and suitable for bodywork up to 45 seats. In 1970 the VAL was replaced by the YRT and the VAM by the YRQ. The VAS continued in more limited production to the end of Bedford in 1986. After 1968 it became, according to the General Motors computer, the PJK. Examples of all three chassis are currently preserved.

Below left:
Twenty nine-seat Willowbrook bodywork was fitted to HGA 984D, a 1966 Bedford VAS1 being operated in 1983 by Taylor & Hudson of Windermere. As usual the Bedford badges are prominent but the bodywork is very austere.

Above:
The Bedford VAL was easily distinguished by its twin steering axles. Most were bodied by Duple or Plaxtons. EHL 472D (West Riding 3) is a VAL14 with Plaxton C52F body and dates from 1966 — seen here at Showbus 1987.

Below:
VLJ 232J originated with Shamrock & Rambler but is now in Lincolnshire with Richard Kime of Folkingham. A VAL70, it has Duple Viceroy C53F bodywork which can be distinguished from Plaxtons by the side window-design, windscreen and radiator grille.

BMMO SOS-types

Built: 1923-39

Bodywork types: Single- and double-deck

Engines: BMMO –	4-cyl	4.344	petrol (all models S to IM4)
	6-cyl	5.986	RR25B petrol (IM6 to ON)
	6-cyl	8.0	Type K oil or AEC 6-cyl 7.7 oil (DON and CON)
	6-cyl	6.373	petrol and from 1938 BMMO 6-cyl 8.028 oil (FEDD)

Wheelbase lengths:	15ft 7½in	(S and FS)
	16ft 7½in	(Q and QL)
	16ft 10¼in	(IM)
	17ft 6⅜in	(ON)
	16ft 3¼in	(FEDD)

Typical bodywork makes:

S-series: Brush, Davidson (charabancs); Q-series: Brush; IM- and ON-series: Short Bros; ON-series: Short Bros, Brush, Duple, English Electric; FEDD-series: Brush, Eastern Coach Works, Metro-Cammell, Short Bros

Variants: S, FS, Q, QL, M, IM, ON, CON, DON, SON, SLR, ONC, DDRE, FEDD

Recognition features:

The early models (S, FS, Q and QL) have tram-style rocker-panels along the lower sides but from the M onwards gently curving bodysides are standard. Up to the M all vehicles had a coal-scuttle type bonnet with four horizontal bars across the radiator mesh. From the IM onwards the radiator was taller and squarer with a thin, vertical divider surmounted by a triangle. On the FEDD a slimmer radiator was used with a full-height divider and more rounded corners. The SON coaches featured a full-front, rounded and sharply raked, with a concealed radiator covered by a large grille, plus centre entrance fitted with a sliding door. On early models the Midland Red, Northern or Trent name is cast into the radiator top tank, models with a triangular-topped radiator carry the BMMO or Trent (TMT) monogram. Earlier full-fronted models carry the Midland Red name in an elliptical box above the windscreen and later coaches the BMMO monogram on the concealed radiator.

General remarks:

The Birmingham & Midland Motor Omnibus Co (BMMO) made its first foray into vehicle production in 1923 when it launched its SOS S-type normal control chassis based on the Tilling-Stevens TS-3 design fitted with bodywork to seat 32 passengers in bus form. A number of these early models were bodied as charabancs. Some of the earliest chassis were actually built by Tillings but with BMMO specifications. A forward control version appeared in 1926 and designated SOS FS. Of these 18 received charabanc bodies but

Left:
HA 2476 was one of the early BMMO SOS S-types (the similar HA 3501 is preserved) built in the company's workshops from 1923-27. The S was based on the Tilling-Stevens TS3 chassis and gave very reliable service with its BMMO modifications. Note the Midland Red name on the radiator, intricate lined-out livery and high floor line. BMMO

Top:
After experience with the normal control S a forward-control design was a logical development. The FS of 1927 was quickly followed by the Q and QL and then, in 1929, the M. HA 4907 was a Ransomes-bodied SOS M and had bus seats for 32 passengers. The M was basically a QL with more comfortable seating. Ian Allan Library

Above:
The SOS SON of 1935 was a vast improvement on earlier BMMO types and provided modern looks and comfort. Trent Motor Traction 417 (RC 7927) is now preserved by the company — it was new in 1940 and has Willowbrook B34F bodywork. Having been sold out of service to a showman it was subsequently repurchased by the company and completely rebuilt.

three years later were rebuilt as buses. In 1927 came the SOS Q which was followed in 1928 by its low-floor variant the SOS QL. Both were developed to obtain the high capacity of the FS and were forward control designs. The SOS M which made its appearance in 1929 used the QL chassis but had improved bodywork and more comfortable seating. The SOS IM, which began life in 1930, had originally been classified SOS COD, and the series comprised the IM4 (four-cylinder engine) and IM6 (six-cylinder engine). Both remained in production until 1934 when the SOS ON was introduced. This took advantage of the new maximum permitted length for single-deckers of 27ft 6in and accommodated 38 seated passengers. A diesel version was offered — the SOS DON — having 36 seats and a larger power unit. The SOS SON of 1935 was the final prewar single-deck bus and was produced until 1940. A number of these petrol-engined buses were converted to diesel and reclassified SOS CON.

On the double-deck front a prototype had appeared in 1931 — the SOS DD-RE, ie 'double-deck rear entrance' — and this led to a production run of 50 in 1932-33. A prototype front entrance vehicle was built in 1933 — the SOS DD-FC — and 335 examples were built before production ceased in 1939, designated FEDD. Many of the early Brush-bodied examples were completely refurbished in the immediate postwar years.

A coach was introduced in 1937 and had a production run of 50. Classified SOS SLR it featured forward control, full-front with concealed radiator and a centre entrance with sliding door. Finally, in 1939, 25 SOS ONC coaches of a basically similar appearance were built but had somewhat plainer bodywork and a sliding roof.

Examples of the S, QL, IM4, CON, DON, SON and FEDD have been preserved.

BMMO S-types

Built: 1946-70

Bodywork type: Single-deck bus

Engines: BMMO – 6-cyl 8.028 oil (S6-S516)
6-cyl 10.5 oil (S17-S23)

Wheelbase lengths: 16ft 0in (S6-S10) — 27ft 6in bodywork
17ft 6in (S12-S13) — 29ft 3in bodywork
16ft 4in (S14-S15) — 30ft 0in bodywork
18ft 6¼in (S16-S23) — 36ft 0in bodywork

Typical bodywork makes:
BMMO, Brush, Metro-Cammell plus BMMO with Plaxton and Willowbrook finishing

Variants: S6 to S23 but designations S7, S18 not used

Recognition features:
All are full-fronted with concealed radiators. The radiator grille on the S6 is of semi-elliptical shape with horizontal slats divided vertically and with a prominent radiator-filler cap. A similar style was used on S8 to S12 but with vertical edges and slats sloping down towards the sides. On S6 to S12 the lower windscreen edge curved downwards towards the sides. Models S13 to S17 have a smaller, neater grille with vertical slats. S14 onwards have a two-piece windscreen with horizontal framing. The S21 has a wider grille incorporating the headlamps and having horizontal slats, and the S22 and S23 have vertically-slatted grilles modified from that used on the double-deck D9. From S21 onwards peaked front domes are used, some S14s have hopper windows and the S21 and S22 have forced-air ventilation and plain windows.

General remarks:
Some 100 single-deckers introduced to the Midland Red fleet in 1946-47 resulted from wartime development work on five prototypes designated S1 to S5. The S6 was built in equal numbers by Brush and Metro-Cammell, had full-fronts and underfloor engines. Originally built to carry 40 seated passengers, all were 'stretched' by Roe in 1953 and fitted with 44 seats. Code S7 was not used but the S8 appeared in 1948-49, again as 100 vehicles, all with 40-seat Metro-Cammell bodywork basically similar to the S6. They

Top:
The BMMO S9 was fairly typical of the earlier S-types although it is not always easy to tell one variant from another. This type of grille was fitted to the S8 to S12 and from S6 to S12 this style of windscreen was fitted. All have the Midland name on the front panel. LHA 410 is a Brush-bodied 1949 example. Ian Allan Library

Above:
The BMMO S14 onwards were much more modern designs with completely redesigned fronts and squarer bodyshells. 651 AHA is an S14 of 1955 — Midland Red 4651. BMMO

---

were also lengthened and reseated by Roe between 1951 and 1953. In 1949 100 S9s followed, basically similar but with Brush bodywork and 8ft 0in wide like the S8 (the S6 was 7ft 6in wide). Delivery of 145 S10s began in 1949, with Brush and Metro-Cammell

Top:
The BMMO S21 of 1967 was mechanically similar to the S17 but with the body structure based on the CM6 coach. The windscreen was also based on the S17 but a full-width grille was fitted. An added refinement was forced-air ventilation with fixed windows. No 5855 (JHA 855E) is seen here in the attractive red and black livery adopted for the series. G. H. F. Atkins

Above:
The BMMO S22 and S23 reverted to the earlier general style as shown by S23 5919 (RHA 919G) which had BMMO B51F bodywork. Now immaculately preserved it was photographed at a Bristol rally. S. C. Morris

bodies, these were also lengthend by Roe. The maximum permitted length for single-deckers had been 27ft 6in but in 1950 the law was amended to allow 30ft 0in thus the 44 S12s of that year were built to 29ft 3in by Brush and Metro-Cammell. (The S11 had comprised a solitary prototype — a S10 with modified suspension — and the S13 was a similar experimental vehicle.) The S13, however, then went into production between 1951 and 1953 as a dual-purpose-seated batch of 98 bodied by Brush and Nudd Bros & Lockyer. The latter had a modified, semi-elliptical, front grille which was used until the S17. Some 218 S14s were built in 1955-59, these being BMMO's first integrally constructed vehicles to be built at Carlyle Works, and weighing just over five tons; they were the first to be built for OPO use. The S15 of 1957 was a dual-purpose variant of the S14 though all were later demoted to bus use. In 1961 the new legal maximum length of 36ft 0in saw production of the S16 of 1961-62, the 85 constructed being in effect a 52-seat lengthened S14. The extra weight caused it to be underpowered so the 10.5-litre engine was substituted on the S17. A further batch of S16s was built in 1964.

Many S17s had the bodywork finished by outside contractors, Plaxton and Willowbrook being chosen. The designation S18 was not used and a solitary modified-suspension S17 became S19. The final series broke with tradition although mechanically similar to the S17. Built to 36ft 0in×8ft 2½in overall the body structure was modelled on the CM6 coach. Thirty dual-purpose, 49-seat S21s entered service in 1967 and were fitted for OPO in 1970. Thirty-seven 45-seat buses, designated S22, followed and these were OPO fitted from new and had radiator grilles similar to those of the D9 double-decker. The final delivery was of 75 S23 OPO service buses in 1968-70 of which the final 50 had bodywork finished by Plaxton.

NB: The S20 designation was used for a batch of Leyland/Willowbrook dual-purpose vehicles in 1966. In December 1973 a total of 14 Class S16 and 79 Class S17 passed to the West Midlands PTE. A good selection of S types have been preserved and include the S6, S12, S14-S17 and S21-S23.

BMMO C-types

Built: 1948-65

Bodywork type: Single-deck coach

Engines: BMMO – 6-cyl 8.028 oil (C1 to C5 series)
6-cyl 10.5 oil (CM6 series)

Wheelbase lengths: 16ft 0in (C1 to C4)
16ft 4in (C5)
18ft 6¼in (CM6T)

Typical bodywork makes:
Duple (C1 and C2), Plaxton (CL2), BMMO (C4, C5 and CM6)

Variants: C1, C2, CL2, C3, CL3, C4, C5, CM5, CM5T, CM6, CM6T

Recognition features:
Classes C1 to C4 have basically similar bodywork with pronounced curves to the side window-line and strongly rounded fronts. The C1 and C2 have four-piece windscreens, twin destination boxes and central doorways — the C1 having a sliding door and the C2 an outwardly hinged door. The C3 and C4 have a modernised body with single piece destination blinds, curved windscreen corner-glasses and two-piece flat windscreens. The central doorways have outwardly hinged doors. A number of Classes C2, C3 and C4 were fitted with Plaxton Embassy bodies in the early 1960s and these are distinguishable by the oval radiator design incorporating the headlamps, large side windows and chromed hubcaps. Radiator covers on the originals all follow BMMO's unique and almost indescribable design. C1 and C2 have two-section fenders whilst C3 and C4 have an 'all-round' type. C4 has roof quarter-lights. The C5 broke new ground with its four-piece continental style windscreen, rectangular grille with rounded corners and chromium-plated bars across. Side windows are straight, the hinged door is at the extreme front and a single line destination display is fitted. The C6 had a six-bay body with revised front and rear, hinged front passenger door, side destination displays and only two bars across the grilles. Some were later fitted with bus-type destination boxes.

General remarks:
The first postwar BMMO coaches were based on the S6 bus chassis and 45 examples of Class C1 were delivered in 1948-49 followed by 12 basically similar C2s in 1950. The C2 was intended for touring work and fitted with 26 seats as opposed to the 30 of the C1. Both classes were bodied by Duple but in 1962 three of the C2s were rebodied by Plaxton and reclassified CL2. The C4 appeared in 1953 with BMMO C32C bodywork and was followed by the Willowbrook C37C bodied C3 in 1954. All the bodywork was to the same basic design. In 1963, 17 C3s were rebodied by Plaxtons and designated CL3; as a result they were then 2ft 0in longer than originally. In 1959 the C4s were upseated to C37F and had their radiator grilles modified.

The C5 was a development of the S14/15 bus having the same mechanical specification and body framing. Produced between 1958 and 1961 it featured several new design features, the most prominent being the continental-style 'lantern' windscreen and forward entrance. The CM5 was a turbocharged variant and the CM5T was fitted with a toilet, the latter type inaugurating Britain's first motorway express service on the newly opened M1 motorway. A further variation was the CS5 for use on normal coaching duties but capable of motorway use. The CM5s totalled 54 and the CM5T only 11 but its success led to the CM6T of 1965 of which 29 were constructed following trials with a prototype built two years earlier. These had semi-automatic gearboxes and a modified windscreen. Of the 29 built, five were CM6s with 46 seats but without toilets for the Birmingham to Worcester service whilst the remainder were C44F plus toilet. In 1972-73 18 were overhauled and restyled at Carlyle Works, fitted with bus destination displays and dual-purpose seating, together with a new white overall livery. They eventually received the standard National Express coach livery. Examples of types C1, CM5T and CM6T have been preserved.

Below:
Now preserved is Duple-bodied BMMO C1, Midland Red 3301 (KHA 301) of 1948. The distinctive design is easily recognised.

Above right:
Also preserved is BMMO C5, Midland Red 4819 (819 HHA) of 1959. Note the continental-style four-piece windscreen, domed roof, window-line and radiator grille. The C5MT was developed from this for motorway express work. Bodywork is for 37 seats and was built in the company's own workshops.

BMMO D-types

Built: 1949-66

Bodywork type: Double-deck bus

Engines: BMMO – 6-cyl 8.028 oil (D5 and D7)
6-cyl 10.5 oil (D9, D10)

Wheelbase lengths: 16ft 3¼in (D5 and D7)
17ft 1½in (D9)
16ft 9ft (D10)

Typical bodywork makes:
Brush (D5), Metro-Cammell (D7), BMMO (D9 — first 305), BMMO/Willowbrook (D9 — last 45), BMMO (D10)

Variants: D5, D5B, D7, D9 and D10

Recognition features:
The D5, D7 and D9 are all halfcab designs with concealed radiators. On the D5 the radiator grille is identical to that on the earlier S series having a distinct downward curve from the centreline and giving a somewhat sad expression to the vehicle. The D7 is fitted with horizontal front framing giving a much tidier appearance with a modified grille retaining the vertical slats and BMMO badge of the D5. The D9 is easily distinguishable by its front overhang and restyled, almost square grille with chromed surround. The grille on the full-fronted, under-floor engined D10 is a hybrid of the D9 and later S-types. The D5 and D7 have four lower-deck window bays, the D9 and D10 five.

General remarks:
The D5 was based on a World War 2 prototype (known as the D1). A total of 100 D5s was built in 1949 with Brush 56-seat open rear-platform bodywork. A further 100, fitted with platform doors, were built in 1950 and designated D5B. In 1953 the lightweight D7 appeared, all 350 examples having Metro-Cammell 'Orion' 58-seat bodywork. In 1958

MIDLAND
D 7
Midland Red
PRIVATE
VHA 403

ROUND LONDON
SIGHTSEEING TOUR
Scotch whisky
3035HA

the 72-seat D9 appeared as an advanced design with self-changing gears, power-steering, independent front suspension and servo-assisted disc brakes. Its notable features were the restyled radiator, front overhang and one-piece windscreen. Of the 350 built by BMMO the final 45 had the bodywork finished by Willowbrook. The final BMMO double-deck venture was the high capacity, underfloor-engined, front entrance D10, of which two prototypes were built in 1960 and 1961. However, the wider availability of Daimler and Leyland rear-engined buses meant that the D10 did not go into production, the company purchasing its requirements from outside manufacturers. In 1973 a large number of D9s was taken over by West Midlands PTE and repainted into the PTE livery, and in 1975 seven D9s were sold to Obsolete Fleet, converted to opentop, and used on London sightseeing tours. One each of Types D5B and D10, and no less than 12 D7s and 15 D9s have been preserved.

Above left:
The BMMO D7 was visually a modified version of the earlier D5 with straight-topped cab window and cleaner body-lines. The BMMO monogram introduced on the D5 was continued on the D7 and D9. VHA 403, a D7 of 1955 with Metro-Cammell 63-seat bodywork, has been preserved and was being rallied at the 1974 Leicester CT open day when this photograph was taken. T. W. Moore

Left:
The distinctive feature of the BMMO D9 was the front overhang, an unusual feature on a halfcab design. First introduced in 1960 several of the last examples were sold to Obsolete Fleet in 1980 for London sightseeing work and converted to opentop. 3035 HA is a 1963 bus which received this treatment, seen here at London Country's Hemel Hempstead open day in 1982. The BMMO monogram had been covered by an LT bullseye.

Above:
The BMMO D10 was an underfloor-engined high capacity double-decker. Two prototypes were built of which the first, 4943 (943 KHA) is preserved at the Midland Bus Museum, Wythall where it was photographed in September 1986. The D10 was not put into production and double-deck requirements were met by purchases of Daimler Fleetlines. Mike Greenwood

Bristol Early types

Built: 1908-37

Bodywork types: Double-deck bus, single-deck bus and coach

Engines:

Bristol –	4-cyl	petrol		X type (C until 1913)
				Y type (C after 1913)
				AW type (W)
				BW type (4-tonner)
				EW type (2-tonner)
				FW type (A)
				GW type (B)
	6-cyl	petrol		HW type (C 3-axle)
				JW type (D, G and J)
				LW type (H)
Gardner –	5-cyl	7.0		5LW type oil (G and J)

Wheelbase lengths (preserved types): 12ft 6in (2-tonner)
(4-tonner)
16ft 0in (B)
17ft 6in (G, H and J)

Typical bodywork makes: Brislington Body Building Works (BBW), Beadle, Brush, Eastern Counties, Eastern Coach Works, United Automobile Services, Weymann

Recognition features:

The 2-tonner is a squat vehicle, usually having full-fronted bodywork, and with a large oval-topped radiator with tapering sides and three chrome bars across the mesh. The earlier 4-tonner has a wide bonnet and radiator, thin mudguards and headlamps mounted on the front bulkhead. The Bristol scroll appears in large script on the 4-tonner's radiator toptank and on a smaller oval plate on the 2-tonner. The B-type radiator is of a squarer design with oval top and bottom and tapering sides, again with the Bristol scroll on an oval plate on the radiator toptank. The B, G, H and J are all forward control models. The G, H and J have narrower, taller radiators with only slightly tapering sides and the Bristol name on a similar oval plate.

General remarks:

The origin of Bristol Commercial Vehicles lay in the Bristol Tramways & Carriage Company Ltd which began PSV construction in 1908 with the lightweight 16-seat C40 type. This was superseded in 1911 by the larger C45 and in 1912 by the C60. Other

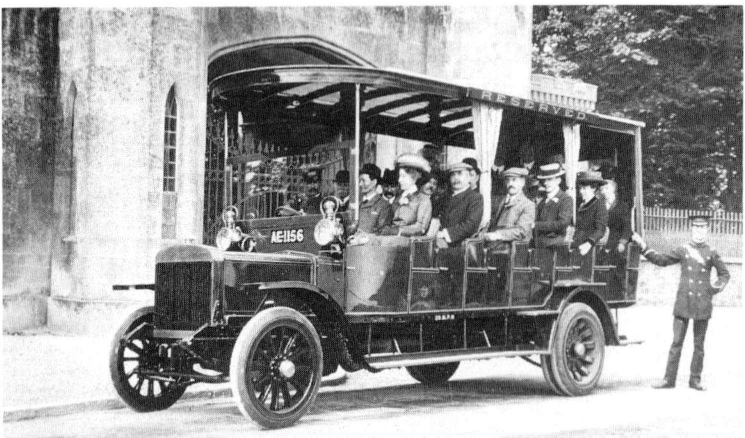

Left:
Typical of the very early Bristol designs was AE 1156, No 73 in the Bristol Tramways fleet, a C45 type with covered charabanc bodywork for 22 passengers. Oil lamps and an exterior horn are fitted and the wide radiator carries no identifying name.
Ian Allan Library

Top:
Bearing the Bristol scroll-type name on its radiator, FB 5890 was one of three A type saloons supplied to Bath Tramways following the display of the prototype (HU 4325) at the 1925 Commercial Motor Show. The radiator is of the style used for all A and most B-types, a destination blind is fitted but a slip-board appears in the cab window. Noticeable are the high-mounted front spotlamps and the folding luggage platform at the rear. Ian Allan Library

Above:
The Bristol G-type and later models used the familiar style of radiator which, on pre war vehicles, was set very high and necessitated small cab windows and windscreen as shown in this view of VV 5696, a preserved JO5G of United Counties. The designation indicates a J-type with Gardner five-cylinder 5LW type oil engine. The early Eastern Coach Works body is of the early style which saw progressive modification over the years whilst remaining instantly identifiable.

C types were the C45 and C50. All these were normal control types powered by Bristol-designed four-cylinder petrol engines of 30 and 40hp. In 1912 production was moved from the aeroplane factory at Filton to new premises at Brislington and chassis of C45, C50 and C60 were soon in full production. All production ceased in 1915 because of war conditions but not before a number of C50s had been supplied to the Royal Naval Air Service. The company at this time provided all its own bodywork.

In 1915 the prototype W-type was built and this led, when production resumed in 1920, to the four-tonner of which 649 were built before production ceased in 1927. In 1923 a smaller two-tonner was introduced. This was Bristol's first forward control design and it was suitable for both goods and passenger applications. More than 250 were built up to 1929. The second forward control design was the A-type of 1925 which was intended for both double and single-deck bodywork. The lightweight 26/32-seat B-type followed in 1926 and was known as the Bristol Superbus, successfully running to 778 examples before production ended in 1934. The B-type became the D-type when a six-cylinder engine was fitted and the C-type was a three-axle double-deck chassis of 1929, of which only two were built and neither of which was bodied.

A new range was announced in 1931 and these set the seal on the future success of the Company. The G-type was designed for double-deck bodywork and achieved a significant step forward in Bristol design work, being of very modern appearance. The H and J types looked very similar, the difference being in the engines, that in the H being a four-cylinder LW unit whilst the J used the six-cylinder JW. Some 248 G-types were built but only 100 of the H, most of which were bought by Western National and United. By far the most successful was the J-type of which over 820 were constructed. The H ceased production in 1934 and the G- and J-types were discontinued in 1937 when the K- and L-types became available.

Bristol K-type

Built: 1937-57

Bodywork type: Double-deck bus

Engines:

Gardner	–	5-cyl	7.0	5LW type oil	Prewar K
AEC	–	6-cyl	7.7	oil	}
Bristol	–	6-cyl	8.15	AVW type oil	} Postwar K
Gardner	–	5-cyl	7.0	5LW type oil	}
		6-cyl	8.4	6LW type oil	
Bristol	–	6-cyl	8.15	AVW type oil	}
Gardner	–	5-cyl	7.0	5LW type oil	} KS and KSW
		6-cyl	8.4	6LW type oil	}

Wheelbase lengths: 16ft 3in (K), 17ft 6in (KS and KSW)

Bodywork makes: ECW, East Lancs, Beadle, Hants & Dorset, Park Royal, Roe, Weymann

Variants: K, KS and KSW each suffixed by a number and letter indicating number of cylinders and engine make

---

Above right:
AJN 825 is a Bristol K5G built in 1939 for Westcliff-on-Sea Motor Services in whose livery it is now preserved. It is fitted with an ECW 55-seat lowbridge body and displays the high-set prewar style radiator. Now owned by Eastern National by which it was operated as No 1269, it is seen at the 1988 North Weald Rally.

Right:
Another 55-seat lowbridge ECW body is borne by Thames Valley Traction 446 (DBL 154), a K6A dating from 1946 and seen at the 1987 Wendover Rally. The lower postwar radiator allowed a much deeper windscreen and carries the Bristol scroll-type name on both top tank and mesh. The postwar ECW bodywork fitted to the K has five lower and seven upper-deck window bays and is 7ft 6in wide.

2A WEST HORNDON
SOUTHEND
'CORONA'
A TASTE TO ENCOURAGE
AJN 825

MAIDENHEAD
CEDWORTH
THAMES VALLEY
DBL 154

Below:
Brighton Hove & District 6447 (HAP 985) is a 1953 KSW6G with highbridge 57-seat ECW bodywork, seen here at a very wet 1987 Showbus. ECW bodywork for the KS and KSW has four lower and six upper-deck window bays and, at 8ft 0in wide, is recognisable by the wider windscreen than that of the K. The bodywork is generally more rounded and bears the characteristics carried over into the later Lodekka and VRT.

Recognition features:
Prewar vehicles have the earlier style Bristol radiator with parallel sides and oval top and base, high mounted with the front frame cross-member clearly visible. All later models have standard oval-topped radiator tapering in towards the slightly curved base, the whole being set much higher and supported on triangular mounting brackets. Headlamps stand clear of bodywork, windows have rounded corners on Eastern Coach Works bodywork. Prewar vehicles sometimes rebuilt to postwar styling with low radiator which can confuse recognition.

General remarks:
Launched in 1937 the K became the standard Bristol double-decker until superseded by the Lodekka in the 1950s, a total of 4,157 being built. The prewar vehicles were fitted with the Gardner 5LW engine. Production was suspended in 1941 but recommenced in 1944 using the Gardner 5LW and AEC 7.7-litre engine. In 1946 the high-profile radiator was replaced with the more familiar tapered design. In 1947 the Gardner 6LW was added to the engine options and in 1948 the new Bristol AVW was introduced. The KS replaced the K in 1950 and allowed for an extra foot length of bodywork. Production of the KS ended the following year but a further six were built in 1957 using Gardner 6LW engines. The 8ft 0in-wide KSW was introduced in 1950 and remained in production until 1957. Production totals were: 842 prewar K, 1,954 postwar K, 231 KS and 1,130 KSW. Preserved examples include K5G, K6A, K6B, KS5G, KS6G, KSW5G, KSW6B and KSW6G variants.

Bristol L-type

Built: 1937-54

Bodywork types: Single-deck bus and coach

Engines:

Gardner –	4-cyl	5.6	4LW type oil	Prewar models
	5-cyl	7.0	5LW type oil	
	6-cyl	8.4	6LW type oil	
AEC –	6-cyl	7.7	oil	Postwar models until 1948
Gardner –	4-cyl	5.6	4LW type oil	
	5-cyl	7.0	5LW type oil	
Bristol –	6-cyl	8.15	AVW type oil	Postwar models from 1948
Gardner –	5-cyl	7.0	5LW type oil	
	6-cyl	8.4	6LW type oil	

Wheelbase lengths: 17ft 6in (L)
19ft 0in (LL and LWL)

Bodywork makes: Most chassis were fitted with Eastern Coach Works bodywork but Burlingham, Duple and Willowbrook bodies were also used

Variants: L (27ft 6in long – 7ft 6in wide) — L4G, L5G, L6G, L6A, L6B
LL (30ft 0in – 7ft 6in) — LL5G, LL6G, LL6B
LWL (30ft 0in – 8ft 0in) — LWL5G, LWL6B

Recognition features:
Prewar models have typical high-set oval top and base Bristol radiator. Postwar models have the lower profile, tapered radiator with triangular supporting brackets. Large hole in radiator front for crank-handle. Headlamps standing clear of bodywork. As with the K, prewar vehicles were sometimes rebuilt to look like postwar models.

General remarks:
Introduced in 1937 for single-deck bodywork most prewar examples were fitted with the Gardner 5LW engine (L5G). When production ceased in 1941 a total of 875 had been

Above:
Western National 333 (DOD 518) looks a fairly typical prewar Bristol L5G and, whilst the bodywork looks to be by ECW it is, in fact, by Beadle of Dartford — a fact given away by the slightly flared skirt and the windscreen profile — and was fitted to the 1939 chassis in 1950. Seating is for 36 with a rear entrance.

SOUTHEND
19 COLCHESTER
MALDON
ONO 49
309

CHELTENHAM
SERVICE
11X
CHELTENHAM
CNH 862
428
UNITED COUNTIES
UNITED COUNTIES

built. In 1942 a further 15 'unfrozen' chassis were constructed. The L made its reappearance in 1946 with the modified lower-profile radiator and a wider choice of engines. In 1948 the Bristol AVW and Gardner 6LW engines were offered whilst at the same time the AEC engine was withdrawn from the options. Production of the L ceased in 1950 and was superseded by the LL designed for the newly permitted 30ft 0in maximum length bodywork. After a short production run the LL was withdrawn in 1951 in favour of the LWL which had been offered as an export model since 1948. The LWL was an 8ft 0in wide version of the LL and was produced until 1954, although few were built after 1952.

Left:
This postwar L5G — Eastern National 309 (ONO 49) — is bodied by ECW and dates from 1950. The windscreen and lower body profile, along with the roofline and window design are ECW hallmarks. The lower postwar radiator is another identifying feature.

Below left:
The L was superseded by the longer LL and then by the wider LWL. United Counties 428 (CNH 862) seen at the 1984 Milton Keynes garage open day is a 1952 LWL5G with ECW DP33R bodywork. The livery style was the standard Tilling Group arrangement for coaches.

Bristol LS and MW Series

Built: 1950-58 (LS), 1957-67 (MW)

Bodywork types: Single-deck bus and coach

Engines:

Gardner –	4-cyl	5.6	4HLW horizontal type oil (LS4G)
	5-cyl	7.0	5HLW horizontal type oil (LS5G, MW5G)
	6-cyl	8.4	6HLW horizontal type oil (LS6G, MW6G)
Bristol –	6-cyl	8.15	AVW type oil (LS6B)

Wheelbase lengths: 16ft 10in (MW)
17ft 2in (LS)

Typical bodywork make: Eastern Coach Works

Recognition features:

Both are forward control models with full-fronts. The LS is easily distinguished by its vertical front and four-piece flat windscreen, high-mounted destination box and single headlamps. A low-mounted foglamp was also fitted. The front panel of the LS usually carried the winged Bristol motif. The MW is distinguished by a vertical lower front panel and sloping windscreen. The lower panel has a rectangular mesh-grill with rounded corners and chromed surround. A four-piece windscreen is fitted and many coach-bodied examples have the offside front windows swept down to meet the bottom of the windscreen.

General remarks:

The Bristol LS, introduced in 1950, is an integrally constructed vehicle with overall dimensions of 30ft 0in long by 8ft 0in wide. The engine is mounted horizontally under the floor midway between the axles and a five-speed synchromesh overdrive gearbox plus vacuum servo braking is standard. 1,409 were constructed, all but 19 of them at Eastern Coach Works. The MW is a heavier, separate chassis vehicle than the LS but provides for bodywork of the same dimensions. Gearbox and braking are identical to the LS but only two engine options were available. The majority, out of a total of 1,965, were fitted with Eastern Coach Works bodywork.

Top:
The Bristol LS bus body by ECW was a very box-like affair with vertical front and only slightly inclined windscreens. The front panel was plain and often carried no identifying badge. An unusual feature, as shown on Lincolnshire 2206 at Scunthorpe Bus Station, was the centrally mounted fog lamp. Many bus-bodied examples were also fitted with dual-purpose seating as can also be seen here. R. H. G. Simpson

Above:
The LS coach body by ECW bore a strong resemblance to the MW but with a more rounded general profile and hinged passenger door. Twin foglamps were also fitted and two single-line destination indicators.

Above:
Preserved Bristol Omnibus 2514 (536 JHU) exemplifies the ECW-bodied Bristol MW series. An MW5G it dates from 1961 and has the usual 45 seats. Most production MWs had this style of front panel with round-cornered radiator grille, Bristol 'wings', four-piece windscreen and T type destination blinds. A single near-side foglamp was also normal.

Bristol SC and SU

Built: 1955-61 (SC), 1960-66 (SU)

Bodywork types: Single-deck bus and coach

Engines:

Gardner –	4-cyl	3.8	4LK type oil	SC 4LK type
Perkins –	6-cyl	4.73	P6 type oil	one SC prototype
Albion –	4-cyl	4.1	EN250 type oil	SU series

Wheelbase lengths: 14ft 4½in (SC), 12ft 6in (SUS), 15ft 0in (SUL)

Bodywork make:
Eastern Coach Works

Recognition features:
The SC is a forward-control design with full front and front-mounted engine. Front axle right forward with doorway behind. Bus version has an oval-shaped radiator grille whilst the coach version has a three-part horizontal grille comprising a centre section with vertical bars superimposed on a larger square design with horizontal bars. Window at nearside front to give better kerb visibility. Separate driver's access door on offside and one on near-side to give access to the engine.

The SU is superficially similar to the bus-bodied MW but has a rounded top to the front panel and a higher-set cooling grille for the concealed radiator. The rear has a much more rounded roof-line in conjunction with two smaller rear side windows and a single rear window. The body height is some 9in lower than that of the MW and with a floor height of only 2ft 10in above ground level longitudinal seating is fitted over front and rear wheel arches. The coach version has higher bodywork and floor level to facilitate forward facing seats throughout.

Above:
ECW bodywork for the Bristol SC4LK was recognisable by its windscreen shape and body lines, and also by the standard Bristol-shape radiator grille flanked by two others of a semi-elliptical shape. The door is behind the front axle and also on the nearside is an access door for the engine compartment. In front of this a low window to give the driver better kerbside visibility. 606 JPU began life with Eastern National (No 429) in 1957 and ended up with Silver Star of Rhosgadfan near Caernarfon. Kevin Lane

Below:
ECW provided a quite different frontal appearance to the SC4LK coach as seen in this view of preserved Lincolnshire Road Car 2611 (RFU 689). Apart from the radiator grille, twin foglamps and slightly rounded top to the front panel the exterior appearance is similar to the bus version. Thirty three dual-purpose seats are fitted to this example which dates from 1958.

Right:
The SC was replaced by the SU in 1960 and in appearance the Eastern Coach Works bodywork was a scaled down MW. Western National 1202 (336 EDV) dates from 1961 and is a coach-bodied version with all forward facing seats (DP33F). Having a 15ft 0in wheelbase and an Albion engine gives it the code SUL4A.

General remarks:
The SC was a lightweight small-capacity single-deck model offered with a Gardner 4LK engine mounted vertically at the front and using a five-speed overdrive gearbox and dry-plate clutch plus vacuum servo type brakes. Almost 300 were built mostly with bus bodywork for rural operations, and supplied mainly to Crosville, Eastern Counties and Lincolnshire Road Car. Seating was generally 33 in the coach version and 35 in the bus. Examples of both types are now preserved.

The SU joined the MW in 1960 as a small underfloor-engined chassis available in two lengths for 30 seats (SUS) and 36 seats (SUL). Like the SC it featured a David Brown gearbox and Austin rear axle. Sales were limited, only 181 being built of which 133 went to the Southern and Western National companies.

Bristol Lodekka

Built: 1949-50 (prototypes), 1953-66

Bodywork type: Double-deck bus

Engines:	Bristol –	6-cyl	8.15	AVW type oil (until 1957)
		6-cyl	8.9	BVW type oil (after 1957)
	Gardner –	5-cyl	7.0	5LW type oil
		6-cyl	8.4	6LW type oil
		6-cyl	10.45	6LX type oil (late-model FLFs)

Wheelbase lengths: 15ft 11in (LDX prototypes 1949-50)
16ft 8½in (LDX prototype of 1958, LD, LDS, FS)
18ft 6in (LDL)
19ft 3in (FL)

Bodywork type:
Eastern Coach Works

Variants: LDX prototypes
LD first production models
LDS short wheelbase, 27ft 0in × 8ft 0in bodywork
LDL long wheelbase, 30ft 0in × 8ft 0in bodywork
FS as LDS but with flat-floor
FSF as FS but with forward entrance
FL as LDL but with flat-floor
FLF as FL but with forward entrance

Recognition features:
The first two prototypes had bodywork similar to the K series but with much wider radiator shells. Early production models had a full-length radiator grille, subsequently superseded by a smaller design with chrome surrounds. This was ultimately developed into a much bolder design with heavier chrome surround and screen. All carried the Bristol name at the top of the radiator, many also carried the Bristol scroll type name on the radiator mesh. Models with the later grille carry the Lodekka name at its base. Vehicle front on rear entrance models has a distinct slope whilst the back is vertical; rounded front and rear top-deck side windows; single-piece flat windscreen; high bonnet with flat front incorporating headlamps and radiator grille.

General remarks:
The prototype Lodekka was produced in 1949 as a revolutionary highbridge type model within lowbridge dimensions. This was achieved by the use of a very low-slung chassis and a drop-centre rear axle, thus eliminating the need for a step up from platform to lower-deck gangway. This in turn allowed a conventional highbridge type central gangway on the upper deck. The two prototypes had mechanical variations and both were fully tested in service before production began. Bristol and Gardner engines were available although sales were limited to the nationalised sector, a fact which led in 1956, to the Lodekka being built under licence by Dennis (the Loline), production of which ran until 1967. The Lodekka was offered with a four- or five-speed gearbox and this was attached directly behind the engine, necessitating a rear-facing bench seat across the front lower-deck bulkhead. Braking was of the triple vacuum servo type though air brakes were offered later. Prototype flat-floor versions were produced in 1958 and this led to the elimination of the sunken lower-deck gangway on all subsequent models. The short wheelbase version was classified LDS and the long wheelbase bus LDL. In 1959 a reclassification took place whereby four F series models were introduced with two or three letter designations. Most successful of the later Lodekkas was the FLF with its forward entrance and 70 seats. In all 5,217 Lodekkas were built comprising LDX prototypes (4), LD (2,179), LDL (6), LDS (8), FL (45), FSF (218), FS (890) and FLF (1,867). Now in preservation are examples of the LD and LDS, FS, FSF, FL and FLF, the most numerous being the LD and FLF. The fourth prototype LDX is also preserved and a number of FLFs have been exported for further use.

Above right:
The original Bristol Lodekka LD introduced the wide bonnet and large radiator. Lincolnshire 2302 (KFW 313) was an ECW H33/25RD-bodied LD6G of 1954 which was operated until 1969. It then saw further service with West Riding and was scrapped in 1971. Similar 2318 (LFW 326), though with a Bristol engine, is now safely in the care of the Lincolnshire Vintage Vehicle Society. R. H. G. Simpson

Right:
The Bristol LDL introduced the smaller radiator grille and, with its longer wheelbase, necessitated an extra half window on both decks. By 1980 former Western National LDL6G VDV 752 had been cut down to opentop and was operating with Eastern Counties at Cromer. The longer bodywork allowed an H37/33RD seating arrangement.

SCUNTHORPE
KIRTON INGHAM
103
Ty·Phoo TEA
St Margaret
HOSIERY & KNITWEAR
2302
KFW 313
HVL 749

SEASIDE SPECIAL
CROMER BUS STN
Y LIMITED
SEASIDE SPECIAL
SHERINGHAM · THE RUNTONS · CROMER · OVERSTRAND
EASTERN COUNTIES
VDV 752

Above:
With the introduction of the FS series in 1959 came a modified radiator grille incorporating a chrome surround, though in other respects little change was made to the bodywork by ECW. These two preserved examples from the United Counties fleet have H33/27RD bodywork but KBD 712D has a Gardner six-cylinder engine (FS6G) whilst 651 EBD has a six-cylinder Bristol BVW (FS6B).

Below:
The FS was developed into the FSF with front doorway and then stretched to become the FLF — recognisable by the half window at the rear of both decks. In fact the FLF was built in much greater numbers than its shorter cousin. MVX 878C is a 1965 FLF6G which originated with Eastern National as its No 2841 and now operates for Hedingham and District Omnibuses as its No L101. The longer wheelbase of the FLF allowed 38 upstairs and 32 downstairs seats.

Chevrolet LQ

Built: 1929-31

Bodywork type: Single-deck

Engine: Chevrolet 'Cast-iron Wonder' – 6-cyl 27hp petrol

Wheelbase length: 10ft 11in

Typical bodywork makes:

Many were bodied locally for the operator, known makes include Allen, Bracebridge, Bush & Twiddy, Layne, Spicer and Waveney. Some were also bodied by General Motors

Recognition features:

Low-set chassis with front mudguards sweeping down to base of bodysides, mudguards also having panels between wheels and bonnet. Bonnet and radiator set back from front of frames with large headlamps mounted on either side. Radiator tapers in towards the top and having the hole for the crank-handle in the bottom plate. The Chevrolet name appears on the oval top-tank. Normal control layout.

General remarks:

Chevrolet chassis were the forerunners of the ubiquitous Bedford W-series of the 1930s. A very popular light chassis suitable for 14 or 16-seat bodywork, it was imported from the USA from around 1928, the LQ model proving by far the most popular of a number of variants. This was a 1½-ton truck chassis having a 15ft 7½in long chassis fitted with the new six-cylinder engine and using a single dry-plate clutch and four-speed gearbox, the wheels being of steel disc type. Many LQs were assembled at the General Motors plant at Antwerp from parts shipped from the USA in CKD form. Production ended in 1931, the same year in which the Bedford 2-ton WHG and WLG made their appearance from the former Vauxhall plant at Luton, GM having taken over Vauxhall Motors in 1925.

Other Chevrolet variants popular with bus operators in the 1920s included the LM, LQ, LT, R, U and X of which an X of 1926 and an LO of 1929 are preserved together with an unusual CAP25 waggonette of 1927. Two LQs (of 1928 and 1929) are also preserved.

Below:
The Lincolnshire Vintage Vehicle Society, based at Whisby Road, Lincoln, looks after TE 8318, thought to be a Chevrolet LO but similar to the once very popular LQ. It has bodywork for 14 passengers by Spicer and dates from 1929, having been supplied new to Jardine of Morecambe. It was on view here at the 1983 South Yorkshire Transport Rally. M. Fowler

Maidstone & District LC1 is a Commer Avenger I of 1951 bodied by Harrington (C16C). It shows the distinctive concealed radiator used on this model and the attractive windscreen and body lines for which Harringtons were noted. The coach is now preserved by M&D and has been restored to its original condition.

Maidstone & District
NKN 650

Commer

Built: 1907-57

Bodywork types: Single and double-deck

Engines: Commer – 4-cyl 22hp petrol (WP3)
6-cyl 22.6hp petrol (Avenger, Centaur)
6-cyl 20.9hp petrol (PN3)
6-cyl 28.6hp petrol or 4-cyl 24.8hp oil (PN4 and PNF4)
6-cyl 4.086 petrol or Perkins 6-cyl 4.73 P6 oil (Commando 17A)
6-cyl 4.75 petrol (Avenger I and Contender)
3.26 2-stroke oil (Contender)

Wheelbase lengths: 10ft 6in and 12ft 9in (Centaur)
12ft 9in (PN3)
13ft 9in (PN4 and PNF4)
15ft 9in (Commando 17A and Avenger I)
15ft 7in (Contender)

Typical bodywork makes:
Commer, Dutfield, Harrington, Park Royal, Plaxton, Reall, Reeves-Kenning, Waveney

Recognition features:
Early vehicles carried the name Commer Car in capital letters on top tank of the radiator. Late models have a horizontally-slatted, sloping design with a slight vee profile. A rectangular, horizontally-slatted, concealed radiator was used on the Avenger I, surmounted by four horizontal chromed bars. The Contender is a modern-style full-front design which, as with all others, carries the Commer name on a small chromed plate mounted on the front panel.

General remarks:
The original company was registered as Commercial Cars Ltd and acquired by Humber Ltd in 1927 when the marque name was shortened from Commer Car to Commer. A 20hp single-deck bus had been introduced before 1910 followed by the WP3 in 1911 — a chain-driven 22hp model which was produced until the outbreak of World War 1. In 1929 passenger vehicles were again introduced with the 2-ton Invader, the 6/7-ton G6 and the 32-seat single-deck and 50-seat double-deck Avenger. The Centaur of 1931 was basically a normal control goods chassis and marketed until 1935 when the N range was introduced. The PN3 saw the introduction of the new V-profile sloping radiator and was quickly followed by the PN4 and its forward control variant the PNF4. These were the first vehicles offered with an oil engine but only stayed in production for two years. Following World War 2 the Commando 17A was built from 1946-49 during which time over 1,300 of these normal control chassis were produced. The Avenger I followed from 1949-52, this being a forward control coach chassis having the engine mounted at 70° to the horizontal over the front axle. The Commer-Harrington Contender of 1952-57 was a chassisless version of the Avenger and having 8ft 0in wide bus bodywork by Harrington.

Above right:
This immaculately restored Commer WP3 has seating for 15 and was originally built in 1913. The Commer Car name is prominent on the radiator with oil lamps, exterior horn and sheet metal mudguards as other interesting features. Ian Allan Library

Right:
Showing the later sloping vee-shaped radiator is this Commer PN3 of 1937 with Waveney B20F bodywork which spent many years in the service of Llandudno UDC working circular tours of the Great Orme. V. H. Darling

F236
COMMERCAR
BM 2356

GREAT ORME
E & ADVENT GOSPEL VAN
161
163
JC 4557

Crossley Prewar Types

Built: 1928-39

Bodywork types: Double-deck bus, single-deck bus and coach

Engines: Crossley – 4-cyl 5.3 petrol (Eagle, Hawk)
6-cyl 6.796 petrol (Six, Arrow, Alpha, Condor)
6-cyl 8.367 oil (Mancunian)
6-cyl 9.12 oil direct injection (Condor 1930-33)
6-cyl 9.12 oil indirect injection (Condor 1933-36)

Wheelbase lengths: 15ft 6in (Hawk)
16ft 7½in (all other models)

Typical bodywork makes:
Brush, Crossley, English Electric, Manchester Corporation, Metro-Cammell, Northern Counties, Roe, Strachans (Acton) Ltd, Walker Bros

Variants: Eagle, Hawk, Six, Arrow, Alpha, Condor, Mancunian

Recognition features:
Parallel-sided radiators with the Crossley motif (a Maltese Cross) on the top tank, and/or the Crossley name similarly placed.

Below:
H. Squires of Ruddington, Nottingham operated this Crossley Six which was fitted with bus bodywork by Willowbrook. The prewar Crossley single-deck chassis was equipped with this wide parallel-sided radiator incorporating the company's motif on the toptank with another on the mesh. G. H. F. Atkins

Right:
Crossley adopted a tapering-sided radiator for the Mancunian of 1934 as shown on this Manchester Corporation example photographed earning its keep in the city centre in 1935. Interesting points are the cab front and windscreen, crank-handle and autovac on the nearside front bulkhead. G. H. F. Atkins

General remarks:
Crossley built a number of buses on early goods chassis, of which a 1916-built Type 8 is now preserved, but the first real bus was first introduced in 1928. This was the Eagle of forward control single-deck design which, in the following year, was fitted with a larger engine and sold as the Crossley Six. Shortly afterwards it was renamed the Alpha. In 1930 a double-deck version was marketed as the Condor and in the same year a smaller normal control chassis known as the Hawk, designed for 26-seat bodywork, came on to the market. In 1934 the Mancunian was introduced although this was still basically the same chassis but having the gearbox directly behind the engine. The single-decker remained the Alpha although, confusingly, both Manchester and Sunderland called single-deckers Mancunians! Transmission was normally via a dry-plate clutch and four-speed sliding mesh gearbox to an underslung worm-drive rear axle. Braking was of the triple servo vacuum type. In 1934 one solitary three-axle Condor was built.

Crossley · Postwar Types

Built: 1946-53
Bodywork types: Double-deck bus, single-deck bus and coach
Engines: Crossley – 6-cyl 8.6 HOE7 type oil
Wheelbase lengths: 16ft 7½in (DD42)
17ft 7in (SD42)
19ft 3in (SD42/9)
Typical bodywork makes:
Brush, Burlingham, Cravens, Crossley, Duple, East Lancs, Massey, Roe, Strachans, Willowbrook, Yeates.
Recognition features:
Forward control models with parallel-sided radiators bearing the Crossley name on the toptank or sometimes just a Maltese cross emblem, though later models did not have this feature, and similar bodywork was also built by Crossley — and others — on other chassis, notably for Manchester and Blackburn. Crossley bodywork is distinguished by the rear-window platform support system with its higher sill in the two rear windows.

General remarks:

Crossley introduced the DD42 in 1946 using the same wheel-base as on the prewar series and offering either a constant mesh or turbo-transmitter gearbox, to which were soon added a synchromesh alternative. The designation 42 was derived from 4 wheels, 2 of which were driven. Suffix numbers were then added to identify the variants, these being 3 (7ft 6in wide), 4 (8ft 0in wide), 5 (revised 7ft 6in wide), 6 (Birmingham type), 7 (engine modified 7ft 6in wide), 8 (engine modified, 8ft 0in wide), 8A (improved suspension). Those fitted with the torque converter had an added T suffix.

Production of the SD42 followed in 1947 using a longer wheelbase and these, too, carried suffix numbers — 3 and 6 (7ft 6in wide), 5 (8ft 0in wide), 7 (engine modified, 7ft 6in wide), 8 (engine modified, 8ft 0in wide). In 1950 two further variants were added by extending the wheelbase to 19ft 3in. These were the SD42/9 at 7ft 6in wide and the SD42/9A at 8ft 0in. Some 750 SD chassis were constructed of which around 100 found foreign buyers. On the home market most were bought by independent coach operators.

In 1948 Crossley was taken over by AEC and for some time in the early 1950s some AEC chassis were available badged as Crossleys, eg the Regal Mk III 9621E was known as the Crossley C9621E.

The final DD42 was built in 1952 and the last SD42 in 1953 after which the development work for the AEC Bridgemaster was carried out at the Stockport factory. The works finally closed in 1957.

Below:
The Crossley DD42 series began production in 1946 and Stockport 321 (EDB 575) was a 1951 Crossley-bodied DD42/7. The clean lines of the postwar body designed by the company are very pleasing and the bus presents a smart and uncluttered appearance. Greater Manchester Transport Museum is the current owner of the vehicle. S. C. Morris

Above:
Chesterfield Corporation No 8 (NRA 708) is a Crossley SD42/7 and is also fitted with a Crossley-built body — seating 35 and with rear entrance. Postwar Crossleys had a tall, parallel-sided radiator with the company's name prominently displayed. This particular vehicle was new in 1949 and is now preserved. Ian Allan Library

Daimler Early Types

Built: 1908-25

Bodywork types: Double and single-deck bus

Engine: Knight – 4-cyl 22.4hp sleeve-valve petrol

Wheelbase lengths: 11ft 6in (CB, CC and CD)
13ft 6in (CK and Y)

Bodywork makes:
Those known include Brush, Daimler and Dodson

Variants: CB, CC, CD, CK, CL, CM and Y, KPL

Recognition features:
Normal control models with solid tyres. Daimler name cast into the radiator top tank. Vertically-fluted radiator — the lasting hallmark of Daimler forward-engined PSVs.

General remarks:
The first purpose-built Daimler bus was exhibited at Olympia in March 1908 — a petrol-electric double-decker for the Gearless Motor Omnibus Co Ltd of London. The revolutionary KPL appeared in 1910 and had two 12hp engines (one for each rear wheel) and four-wheel braking. Having secured the services of Frank Searle, Chief Engineer to the LGOC, the next model bore a striking resemblance to the LGOC B-type. Immediately prior to World War I the CB-, CC- and CD-types were introduced. Production continued throughout the war and the wartime version of the CB was classified Y, remaining in production until 1919. The CK used a new four-cylinder 5.1-litre engine (also fitted to the CL and CM of 1925) and was available in either double or single-deck forms. From 1920 pneumatic tyres were offered as an option. In all these early models the drive was via a chain-driven gearbox, cone clutch and worm-drive rear axle. Represented in preservation are an unidentified chassis of 1912, three Y-types and a CK.

Daimler Prewar Types

Built: 1926-40

Bodywork types: Single-deck bus and coach, double-deck bus

Engines:

Daimler –	6cyl	5.76	sleeve-valve petrol (Types CF, CG, CH)
	6-cyl	6.56	poppet-valve petrol (Type CP)
Gardner –	5-cyl	7.0	5LW Type oil (Type COG5)
	6-cyl	8.4	6LW Type oil (Type COG6)
AEC –	6-cyl	7.7	oil (Type COA6)

Wheelbase lengths: 16ft 0in (CF, CG, CH, CP)
16ft 3 5/32in (CO series double-deck)

Bodywork Makes: 17ft 6in (COG5/40 single-deck)
Brush, Dickson, Duple, English Electric, Metro-Cammell, Northern Counties, Roe, Strachans, Weymann, Willowbrook

Variants: CF, CG, CH, CP, COG5, COG6, COA6, COS4 and COT4

Recognition features:
The CF6 was produced in both forward and normal control versions, the CG, CH and CP all being forward-control as are all the CO series vehicles. Radiators are straight-sided and round-topped with a vertically-fluted top tank and, apart from some later COGs, are inclined back from the vertical. The Daimler name is usually displayed on the top tank.

Below:
Typical of the Daimler C series is this preserved Birmingham Corporation COG5 with Metro-Cammell H30/24R body seen at the North Cheshire Cavalcade in May 1977. The radiator, which carries no maker's name, is of the standard Daimler type with fluted top tank. M. Haddon

General remarks:
In June 1926 the Associated Daimler Company was formed by the link with AEC (see separate entry). The Association was dissolved in July 1928 but Daimler continued to produce the ADC 423 and 424 until 1929 when the new 5.76-litre engine was installed and the type became known as CF6. In 1930 it was replaced with the CG6 using the same engine and transmission in a low-frame chassis, but this soon became the CH6 with the installation of a preselector gearbox. The sleeve-valve engine was not popular with operators so the poppet-valve-engined CP6 was introduced in 1931. By this time the diesel engine was growing in popularity and Daimler turned to Gardner which had introduced its LW range of engines that year. The 7-litre 5LW unit was chosen leading to the production of the COG5 (Commercial Oil Gardner Five-cylinder) in 1933. From 1934 double and single-deck chassis were differentiated by the suffix letters DD or SD. The following year the 6LW was used and the COG6 continued in production alongside the COG5 until 1940. Production of the petrol-engined CP6 had ceased in 1935. Coventry Corporation specified the AEC 7.7 litre engine for its orders and COA6s were supplied between 1934 and 1940, all being COA6DD models. A less well known variant was a batch of 10 COS4s supplied to Newcastle Corporation in 1935 and fitted with Armstrong-Saurer four-cylinder engines. In 1934 a unique COT4 fitted with a Tangye four-cylinder oil engine was supplied to Edinburgh Corporation but converted to COG5 two years later. The single-deck COG5/40 introduced in 1936 was designed to allow for up to 40 seats in a 27ft 6in long single-deck chassis and in 1939 a double-deck equivalent, the COG5/60 was produced for Coventry. A COG6/40 with 18ft 9in wheelbase was produced for the export market and three-axle COG5s were built for China in 1940 but sunk in transit. The only three-axle COG6 for the home market was destroyed in an air raid on Coventry in 1940 before it could be delivered to Leicester Corporation. A total of 1,900 CO series double-deckers were built and 400 COG5/40 single-deckers. Vehicles of COG5 and a solitary COS4 are represented in preservation.

Daimler: W for Wartime and V for Victory

Built: 1943-47 and 1946-68

Bodywork types: Single- and double-deck

Engines:

Gardner 5LW	–	5-cyl	7.0	5LW Type oil
Gardner 6LW	–	6-cyl	8.4	6LW Type oil (6HLW in the Freeline)
Gardner 6LX	–	6-cyl	10.45	6LX Type oil
AEC	–	6-cyl	7.7	oil
Daimler CD5	–	6-cyl	8.6	oil
CD650	–	6-cyl	10.6	oil

Wheelbase lengths:

16ft 3⁵⁄₃₂in	CW series
16ft 3⁵⁄₃₂in	CV series
16ft 4in	CV series after 1953 (standard from 1954)
18ft 6in	CV series after 1956 for 30ft 0in bodywork
17ft 2½in	CV series single-deck
16ft 4in	CD650, CC and CL series/Freeline
17ft 6in	Freeline

Typical bodywork makes:
CW series: Brush, Duple, Massey, Northern Counties, Park Royal, Roe and Weymann; CV series: Alexander, Brockhouse, Brush, Burlingham, Crossley, Duple, East Lancs, Harkness, Heaver, Massey, Metro-Cammell, Neepsend, Northern Counties, Park Royal, Roberts, Roe, Strachans, Welsh Metal Industries, Weymann and Willowbrook

Recognition features:
Wartime W series models generally fitted with MoD design 'utility' bodywork of very angular appearance. All exposed radiator models have straight-sided, slightly raked radiators with a vertically-fluted top tank, the CD650 being distinguishable by the increased width of its radiator. Concealed radiator buses have a curved chrome strip with vertical fluting above the ventilation slots in the front panel. CW models had a

Above:
Huddersfield Joint Omnibus Services bought CCX 777 in 1945 and the bus is now safely preserved with regular rally appearances. A Daimler CWA6 it has Duple utility lowbridge 55-seat bodywork of a very basic and angular specification. L. J. Long

Below:
This official photograph shows the first Daimler CD650 of 1949 after being bodied by Charles Roe of Leeds. Some 56 seats were provided in 8ft 0in wide bodywork and the 10.6-litre Daimler engine was used. Daimler

Right:
Also with Roe bodywork is BNH 250C, a 59-seat Daimler CVG6 new in 1965 to Northampton Corporation and preserved by the Rushden Historic Vehicle Trust. The concealed radiator and rounded bonnet is typical of later Daimlers.

coarse mesh grille. Concealed radiators became standard during the 1950s, with a narrower bonnet from around 1957 with headlights outside the bonnet assembly.

General remarks:

The first wartime model to appear from Daimler was the CWG5, a modified COG5, in 1943. The CWA6, a modified COA6, appeared later that year. The W in the chassis designation indicating wartime production. Daimler had undertaken prewar development work on its own diesel engine and the first of these were installed in 1944-built CWD6s. In all some 1,400 wartime specification chassis were supplied to UK operators.

The Victory range was introduced after the cessation of hostilities, first appearing in 1946 as the CVA6, CVD6 and CVG5, the CVG6 following soon afterwards. Using the new 10.6-litre Daimler engine the CD650 was announced in 1948 and for the first time introduced power-steering on a British bus. All these models had preselector gearboxes as standard.

A single-deck chassis was also offered, basically similar to its double-deck counterparts, but stretched to 17ft 2in wheelbase. Introduced in 1947, production continued until 1955. In 1951 Daimler unveiled its first named bus chassis, the single-deck 'Freeline', an underfloor horizontally-engined vehicle designated D650HS or G650HS depending on whether the Daimler horizontal version of the CD650 or the Gardner 6HLW engine was used. Production of the 'Freeline' continued until 1964.

A lightweight version of the CV was introduced in 1952 as the CLG5 weighing only 6ton 2cwt; a CLG6 and CLD6 were soon added but the main weight saving features were soon incorporated into the standard CV range.

In 1956 when the Construction and Use Regulations were revised to permit 30ft 0in-long double-deckers on two axles an 18ft 0in-wheelbase CV was introduced — the CVG6-30. From 1958 the new Gardner 6LX engine was offered, leading to the CVD650-30 and the CVG6LX-30, the latter being by far the most popular with operators.

In 1958 a manual gearbox was offered as an alternative to the preselector thus spawning the CSD6, CSG5 and CSG6 (S indicating synchromesh gearbox). David Brown four-speed gearboxes were used.

The CCG5 and CCG6 superseded these in 1963, having a constant-mesh gearbox. The last five-cylinder bus produced for the UK market was a CCG5 of 1968 for Burton-upon-Trent Corporation and the last home-market front-engined Daimlers were five CVG6s for Northampton Corporation delivered in the same year.

Dennis Early Types

Built: 1904-25

Bodywork types: Double- and single-deck

Engines: Aster – 4-cyl 28hp petrol
White & Poppe – 4-cyl 6.24 petrol

Wheelbase lengths: 13ft 0in to 15ft 0in

Typical bodywork makes:
Dennis, Charlesworth, Dodson, Lewis, Northern Counties, Short, Strachan & Brown

Recognition features:
All early Dennis buses were of normal-control design and had a large square radiator with a very rounded top tank on which was mounted the Dennis name on a cast oval plate. The bonnet sides incorporated a series of vertical ventilation louvres.

General remarks:
The first Dennis bus appeared in 1904 as a 28hp Aster-engined chassis carrying a horsebus body and was used on a service from Kingston to Richmond-upon-Thames. The chassis was suitable for goods vehicles, double- or single-deck bus bodywork. In 1913 the A type of 3-4 tons was introduced, known as the 'Subsidy Model' as customers were given an annual payment for making them available to the War Office on demand. Over 7,000 were built before being superseded by a 40hp chassis in 1919. This was of 14ft 1½in wheelbase allowing for 7ft 0in by 21ft 10in charabanc bodywork which Dennis also supplied. Between 1916 and 1922 Dennis-Stevens petrol-electric engines were also available. The 'Portland' of 1923 was available for 20/25-seat bus bodywork as well as goods applications. Weighing 2-2½ tons it was powered by a four-cylinder 21bhp engine and four-speed gearbox, the 13ft 0in wheelbase allowing for 6ft 6½in by 19ft 8in bus bodywork.

In 1925 the 4-ton 14ft 0in wheelbase chassis was made available for coach bodies seating up to 35 passengers. Those operated by Greyhound of Bristol with 24 seats inaugurated the then longest route in Britain from Bristol to London, a journey of 8hr. The 15ft 10in wheelbase version was available for double-deck buses and was entitled the '40/70hp Omnibus Chassis'. It could seat up to 54 passengers within an overall size of 6ft 10in by 25ft 0in. A smaller 16/22-seat 25/36hp chassis was also marketed in the same year. Interesting survivors from this era are a 1916 2-ton lorry carrying a replica charabanc body (BK 2136) and a 1925 4-ton chassis fitted with a Dodson O48R body (LGOC D142 [XX 9591]).

Above:
This Dennis long-wheelbase four-ton chassis was one of a number which came into the LGOC fleet from Independents. The forward control chassis has the earlier Dennis oval plate on the radiator toptank and is fitted with Dodson 48-seat opentop bodywork conforming to Metropolitan Police regulations in force at its time of construction in 1926. D. M. Stuttard

Dennis Prewar Single-Deckers

Built: 1926-39

Bodywork types: Bus and coach

Engines:

Dennis –	4-cyl	2.72	petrol (G series)
	4-cyl	3.77	petrol (Ace and Mace)
	4-cyl	5.7	petrol (1931 Lancet)
	4-cyl	5.702	petrol (E series and Lancet I)
	4-cyl	6.8	petrol (1936 Lancet)
	6-cyl	3.77	petrol (Dart, 1938 Lancet, Falcon)
	6-cyl	6.13	petrol (Arrow)
Perkins –	4-cyl	3.15	P4 type oil (1935 Lancet)
Gardner –	4-cyl	3.8	4LK type oil (1935 Lancet)
	5-cyl	7.0	5LW type oil (1936 Lancet)

Wheelbase lengths:

10ft 9in	(G series)
11ft 6in	(Ace and Mace, GL)
12ft 6in	(Pike)
12ft 7in	(GL)
13ft 2½in	(Dart)
14ft 0in	(Falcon)
14ft 6in	(Arrow Minor)
16ft 0in	(E series)
16ft 4in	(Lancet)
16ft 6in	(Falcon)
16ft 7½in	(F series)

Typical bodywork makes:
Brush, Dennis, Duple, Eastern Coach Works, Harkness, Harrington, Mumford, Park Royal, Strachans (Acton) Ltd, Thurgood, Weymann, Willowbrook (plus many local builders)

Recognition features:
A variety of radiator designs with louvred bonnet side-panels. Earlier models have parallel-sided radiators with rounded top tanks bearing a cast oval Dennis nameplate. A new round-topped design with tapering sides and curved bottom panel was introduced for the Arrow and used on earlier Lancets. In 1935 a shallower radiator shell with a much greater taper on both radiator casing and bonnet sides became the standard — radiator sides taper towards the base, bonnet top slopes forward, bonnet sides taper towards radiator, Dennis name in smaller letters.

General remarks:
The first true Dennis bus chassis, the E, appeared late in 1925 as a semi-forward control design with a flat-topped chassis with a floor height 7in lower than that on the four-tonner. Pneumatic tyres were standard and the power unit was a modified 40/70bhp engine. The normal control equivalent was classified F. The EV, a modified E, was introduced in 1928 with a normal control FS equivalent. These used the new style Dennis nameplate — a stylised D with 'tails' embracing the other letters of the name. The previous year had seen the introduction of the G, a lightweight chassis for 20-seat bodywork. The GL followed in 1929 along with the Dart using a new six-cylinder engine flexibly mounted on the chassis frame with the four-speed gearbox. The G was discontinued in 1931 and the GL and Dart in 1933.

In 1930 the company developed a new six-cylinder engine for the new Arrow chassis, offered in forward and normal control versions, of which only 58 were built before production ended in 1934. In 1931 the Lancet was announced, suitable for maximum length bodywork and powered by an improved four-cylinder 5.7-litre engine. It sold for £400 less than the Arrow and soon became very popular. The six-cylinder engine of the Arrow was also available thus creating the Lancet 6. In 1933 two new chassis using a larger 3.77-litre engine replaced the GL and Dart. The Ace was suitable for 20-seat bodywork and, with its set-back front axle and distinctive appearance, was soon known as the 'flying pig'! The similar Mace was 30in longer overall and of forward control design to allow for 26 seats.

A new Lancet appeared for the 1935 Commercial Motor Show and used a four-cylinder 6.8-litre petrol or 6.5-litre Lanova type oil engine. In 1936 the Gardner 5LW became available. Two 1937 lorry chassis were made available for bus use, the normal

Left:
The Dennis GL of 1929 was bodied in bus and touring forms and CC 9305, now preserved, carries Roberts 15-seat toastrack body with folding roof. It was new to Llandudno UDC but is now owned by Glenton Tours and seen here at a sunny Showbus rally. Ian Allan Library

Top:
The semi-forward control Dennis Ace had a distinctive tapering bonnet and front overhang as shown by this official photograph. The AD 1934 registration is presumably fictitious although the photograph is dated 1933. The price of £660 included curtains and a sunshine roof! Dennis Bros

Above:
The forward control Dennis Lancet I presented a very different appearance from previous Dennis designs with a radiator shell that was to become more or less standard for many years and on which the Dennis and model name featured. Belfast Corporation bought CZ 7013 in 1935 and Harkness provided the B13F body. It is now preserved by the Belfast Transport Museum. Ian Allan Library

control Pike based on the Ajax 2-ton chassis, and the light 4-ton-based Arrow Minor. They were 20- and 26-seaters but only a few were sold. In 1938 the whole passenger range was rationalised into one model, the Falcon, using the four-cylinder, 3.77-litre petrol engine, four-speed gearbox and spiral bevel rear axle. The Ace radiator was used with a forward control layout although a few normal control versions were also built. Production came to an end with the outbreak of hostilities in 1939.

Examples preserved include the E, G, GL, Dart, Lancet, Ace and Mace.

Dennis Postwar Single-Deckers

Built: 1945-67

Bodywork type: Bus and coach

Engines:

Dennis –	4-cyl	3.77	petrol (Falcon II)
	4-cyl	3.88	petrol (Pax)
	6-cyl	5.07	oil (Falcon II)
	6-cyl	5.5	oil (Falcon II and Pelican)
	6-cyl	7.6	oil (Lancet J3, J10, UF, Dominant)
Gardner –	4-cyl	3.8	4LK and 5.6 4LW type oil (Falcon II)
Perkins –	4-cyl	3.15	P4 type oil (Heron)
	6-cyl	4.73	P6 type oil (Falcon II, Triton, Stork, Pax V)

Wheelbase lengths: 9ft 6in (Triton)
11ft 6in (Pax)
12ft 6in (Stork)
12ft 9in (Pax)
16ft 4in (Lancet UF)
16ft 6in (Falcon II)
16ft 9in (Dominant)
17ft 1in (Falcon II)
17ft 5in (Lancet J3)
18ft 6in (Lancet J10)

Typical bodywork makes:
Barnaby, Brush, Burlingham, Dennis, Duple, Harrington, Plaxton, Reading, Strachans, Weymann, Windover, Yeates, plus many local coachbuilders

Recognition features:
The postwar radiator was a distinctive taper-sided design with heavily rounded top and bottom, the filler cap being set back and the Dennis name appearing on a small plate at the top. The bonnet cover was usually plain but some carried three horizontal chrome strips. Some full-fronted bodywork incorporated the standard radiator but many varied according to the manufacturer's design; most carry the Dennis nameplate below the windscreen.

Below left:
LOU 76 was one of 15 Dennis Falcons delivered to Aldershot and District in 1954 with bodywork supplied by Strachans. Seating was for 30 passengers and the power unit a Gardner 4LK engine. Eight more of this attractive design were purchased in 1956 of which seven were sold two years later to Isle of Man Road Services one of which, 29 HMN (formerly POR 428) has been preserved. Strachans

Top:
The Lancet was developed after World War 2 as the J3 and this Reading-bodied coach was bought by Safeway of South Petherton in 1949. After being out of service for many years it was restored by the company in 1985. The modified Dennis badge is used on the radiator which has a more rounded top than its prewar counterpart.

Above:
Preserved but used on stage work in Barmouth in 1987 is Strachan-bodied Dennis Lancet J10 HOU 904 of 1950. The 38-seat bus bodywork gives a quite different appearance from that of the coach. It was originally Aldershot and District 178 in whose livery it has been restored. G. H. Truran

General remarks:
The Lancet J3 appeared in 1945 and carried a Dennis 32-seat rear entrance bus body, the chassis providing for 29ft 6in bodywork. The J3A was a full-fronted variant. In 1947 the unique J4 was produced for Lancashire United Transport and powered by the Gardner 5LW engine. The 1950 changes to the Construction and Use Regulations saw the J10 (halfcab) and J10A (full-front) for 7ft 6in wide bodywork, and the J10B (halfcab) and J10C (full-front) for bodywork 8ft 0in wide. A number of export Lancet IVs were marketed in 1948 as the J7 (Dennis engine), J7A (Gardner 6LW) and J7B (Dennis 8-litre oil). A left-hand drive Lancet III was also available for a short time coded J5 (halfcab) and J5A (full-front). In 1948 Dennis also attempted to meet the demand for a light chassis suitable for up to 26-seat bodywork by reintroducing the Falcon in both normal and forward-control versions.

At the 1950 Show, Dennis exhibited the Dominant, an underfloor-engined design incorporating a frame rising over the offside-mounted engine. Turbocharging and semi-automatic transmission were also added but only two were built. In 1952 the more conventional Lancet UF appeared using a five-speed gearbox and pre-selector overdrive plus vacuum assisted hydraulic braking. A total of 71 was built before production ceased in 1961. Dennis's answer to continuing demands from operators for lightweight chassis to reduce running costs was the Pelican of 1956 which used some Falcon components but able to seat 44 with a weight of only 5ton 15½ cwt. Only one was ever built but saw almost 20 years of service with three operators. Four other lightweight chassis were also produced based on the postwar goods range. A number of normal control Pax were fitted with coach bodies and a modified version, known as the Triton, was built for the London County Council (LCC) between 1952 and 1958. In 1956 LCC took a PSV version of the Stork, 3ft longer than the Triton; a few Storks were also built for export. A forward control chassis, the Heron, joined the LCC fleet in 1960-63. Equipped with wheelchair lifts, they seated 11 passengers. In 1967 two Pax Vs were supplied to Llandudno UDC and were the last Dennis lightweight passenger chassis to be constructed.

Examples of the Lancet J3, J10 and UF, Falcon II and Pax have been preserved.

Dennis Double-Deckers

Built: 1927-67

Bodywork type: Bus

Engines:

Dennis –	4-cyl	5.7	petrol (H, HS and HV, Lance II)
	4-cyl	6.13	petrol (Lance)
	4-cyl	6.8	petrol (Lance II)
	6-cyl	7.6	oil (Lance K3)
Armstrong-Saurer –	6-cyl	8.55	oil (Lance)
Gardner –	5-cyl	7.0	5LW type oil (Lance K4)
	6-cyl	8.4	6LW type oil (Lance K2, Loline I, II and III)
	6-cyl	10.45	6LX type oil (Loline II and III)
Leyland –	6-cyl	9.8	O.600 type oil (Loline I, II and III)

Wheelbase lengths:
16ft 0in (H series)
16ft 6in (Lance [prewar])
16ft 3⁵⁄₁₆ (Lance K2 and K3)
16ft 4in (Lance K4)
18ft 6in (Loline I)
16ft 8½in and 19ft 0in (Loline II)
16ft 11½in, 18ft 0in and 19ft 2¼in (Loline III)

Typical bodywork makes:
H series: Dodson, Hall Lewis, Massey, Ransomes, Roe, Short Bros
Prewar Lance: Beadle, Birch, Park Royal, Roe, Strachans, Thomas Tilling
Postwar Lance: East Lancs, Strachans, Weymann
Loline: Alexander, East Lancs, Northern Counties, Willowbrook

Recognition features:
The H series had large squat radiators with the Dennis name on a rectangular plate on the top tank and sometimes on the bodywork. All were of semi-forward control and usually had an open cab and overhanging canopy, some had 'piano-front' bodywork. The Lance followed the radiator style described in the section on prewar single-deckers and usually had steeply inclined windscreens. The postwar Lance has a modified radiator with the Dennis name appearing on the mesh, but the K4 has a concealed radiator with 'new-look' bonnet typical of its time and incorporating the headlamps. The Loline I and II have a wide bonnet with 'fish-mouth' grill. The Mk I has rear and the Mk II front-entrance bodywork. The Loline III has a rectangular mesh radiator grille bearing the words 'Loline III' and normally forward entrance bodywork. A lowbridge Loline II with full-fronted, front entrance, bodywork with wrapround screens was supplied to Barton Transport, Chilwell, it was then the lowest closed-top double-decker ever built, at a height of 12ft 5½in.

General remarks:
Appearing in 1927 the H-type used the chassis and engine from the E. The HS had a modified engine and gearbox plus a sturdier rear axle. The HV had the petrol tank mounted in a safer position — under the chassis floor away from the engine. In 1930 the HV was replaced by the Lance, a double-deck version of the Arrow, employing a straight drive-line to an underslung worm differential on the nearside of the rear axle. Just over a year later the modified Lance II with a completely redesigned frame was introduced. Production of the Lance ceased with the outbreak of hostilities in 1939.

Above:
The Lance K3 of 1947 was the Dennis postwar standard double-deck chassis of which GOU 845 was purchased by Aldershot & District (No 145) in 1950. Bodywork is by East Lancs. A new radiator having a more oval top, slightly tapering sides and vee-shaped base was used and this carried the Dennis and Lance names. 145 is seen at the 1987 Showbus rally.

Production resumed in 1947 with 10 Gardner-powered buses for Lancashire United Transport using a chassis very similar to the prewar Lancet II, but having a redesigned radiator shell and bonnet. The model was known as the Lance K2. The Lance K3 of 1947 became the standard production chassis, being basically similar to the K2. The K4 of 1954 allowed for 27ft 0in bodywork, the main differences between the three being embodied in the engines. Lance production ended in 1954.

Bristol's successful Lodekka was restricted to nationalised operators, so Dennis began to build the chassis under licence in 1956, thus making it more widely available. The Loline I incorporated a Dennis front axle, double-plate clutch and constant-mesh overdrive gearbox. It was designed for 30ft 0in bodywork. The Loline II of 1958 was an improved version with wider chassis sides for greater stability and designed for front entrance bodywork to enable the driver to supervise loading. A choice of chassis lengths was available together with three different engines. Air suspension on the rear axle was standard. In 1960 the improved Loline III appeared in which the chassis frame allowed a flatter floor and the fuel tank was set behind the rear axle. Four or five-speed semi-automatic gearboxes were available and a choice of engines in three different length wheelbases. Production ceased in 1967.

Examples of the Lance K3 and K4 and the Loline I, II and III have been preserved.

Left:
Dennis is probably best known for the Loline which went into production in 1956 and of which Aldershot and District 357 (SOU 465) is a good example. New in 1958 it passed to Alder Valley before preservation. Bodywork is by East Lancs and the A&D name rather than that of Dennis is carried on the front cowl. Ian Allan Library

Below:
EPG 179B was also new to Aldershot and District but has Northern Counties 75-seat bodywork. A Loline III it was new in 1964 and passed to Warner's of Tewkesbury and then Provincial and Strand in Bedfordshire by which time the destination boxes had been panelled over. It has the normal Mk III concealed radiator and front sliding door.

Dodge

Dodge Bros began to assemble vehicles at its Kew factory in 1927 using engines and gearboxes made in the USA. In 1931 a new range of 1½ and 2-ton normal control trucks was introduced and many of these were bodied for PSV use. One of the 1½-ton, UF30A series (registration VK 5401) is now preserved. The range was powered by a Dodge 21hp four-cylinder sidevalve petrol engine driving through a four-speed gearbox and single-plate clutch to a spiral-bevel rear axle. The wheelbase is 11ft 4in long and the Robson bodywork on the preserved example seats 14. The company merged with Rootes in 1964 and in 1967 all production was transferred to the Commer/Karrier factory at Dunstable. In 1976 the Commer name was discontinued, all subsequent truck models carrying either the Dodge or Karrier name. Since 1981 the operation has been part of the Renault group.

Foden

Built: 1945-56

Bodywork types: Double and single-deck bus, single-deck coach

Engines: Foden – 4.09 FD6 type 2-stroke oil
Gardner – 5-cyl 7.0 5LW type oil
6-cyl 8.4 6LW type oil

Wheelbase lengths: 16ft 3in (PVD)
16ft 6in (PVR)
17ft 6in (PVSC)

Typical bodywork makes:
Crossley, East Lancs, Lawton, Massey, Plaxton, Metalcraft

Recognition features:
Variations on the vee-design radiator grille, the PVD and PVSC being of halfcab design and the PVR having a full-front with 'Foden' in a shallow diamond design. All postwar front-engined chassis had concealed radiators and full-width bonnets with a pronounced curve downwards towards the nearside on half cabs.

General remarks:
The firm of Hancock & Foden of Sandbach became established as builders of steam wagons, moving into diesel lorries in 1932. In 1945 the DG range of passenger models was announced. The production version was known as the PVD, a double-deck design with full-width bonnet and concealed radiator, offered with either the Gardner 5LW or 6LW engine (PVD5 or PVD6). In 1948 the Foden FD6 engine was added although only one bus was ever fitted with it. When production ended in 1956 less than 50 PVDs had been sold. In 1946 the PVSC, a vertically mounted front-engined single-deck chassis was offered with the 5LW or 6LW engine (PVSC5 or PVSC6). As with the PVD the Foden FD6 became an option and was classified PVFE6. A rear-engined coach was introduced in 1950 known as the PVR and could be had with the Foden FD6 (PVRF6) or Gardner 6LW engine (PVRG6). Production of this ceased in 1955 and the PVSC in early 1956. Examples of all three series have been preserved.

Below left:
Front-engined Fodens were distinguishable by the use of variations on the company's standard concealed radiator design. The double-deck PVD had this almost symmetrical curve to the bonnet top and lower cab windscreen making a very attractive vehicle. This 1955 example is seen before delivery. Ian Allan Library

Above:
Foden PVSC6 LMA 284 has fortunately been preserved and can be found at the North West Transport Museum, Burtonwood. It has unusual Lawton C35F bodywork and dates from 1949. S. C. Morris

Below:
Only one Foden PVR has been preserved — NTU 125 — and is a PVRF6 with Metalcraft C41C bodywork which spent its whole working life with Hollinsheads Coaches in Cheshire. It dates from 1951 and the body design is typical of the period. The rear-engine is identifiable by the grille on the rear offside. C. L. Caddy

Ford

Built: 1908-35

Bodywork type: Single-deck bus

Engines: Ford – 4-cyl 22hp petrol (T and TT)
4-cyl 14.9hp petrol (AA and AF)
4-cyl 24hp petrol (BB) — Ford 3.6 30hp V8 from 1933

Wheelbase lengths: 8ft 4in — T
10ft 4in — TT
8ft 7½in — AA and AF
10ft 11½in — BB
13ft 1in — BB

Typical bodywork makes:
A wide variety of local coachbuilders were employed and many earlier vehicles had interchangeable goods and passenger bodies

Recognition features:
T — parallel-sided radiator with large top tank and filler, dished frame below radiator base, low bonnet and spoked wheels. The TT differed in having a wider, round-topped radiator and bonnet. The AF and BB had a sharply tapering bonnet. All of normal control design with high floor levels.

General remarks:
The world-famous Model T Ford first appeared in 1908 and within 10 years over a million had been sold. Although at first imported from the USA in left-hand drive form, production commenced at Manchester in 1911. The TT, introduced in 1917, was simply a longer wheelbase version. Production of both ceased in 1927 with the introduction of the AF. An improvement over the T and TT was the AF's three-speed gearbox in place of the earlier two-speed. In 1928 the AA, an improved AF with four-speed gearbox began production. This was replaced by the BB in 1932 using the new V8 engine from the

Below:
The Model T Ford was the chassis which got so many transport undertakings on the road after World War 1. One such was Silver Queen of Lincoln whose 47 vehicles — mostly Model Ts — were taken over by Lincolnshire Road Car on its formation in 1928. Several are seen in this photograph at the Bracebridge Heath depot along with a Thornycroft charabanc. They were bodied by Thompson of Louth, following the usual practice of employing a local bodybuilder, with seating for 14. F. G. Moss

following year and offered with two alternative wheelbase lengths. Production ceased in 1935. In 1932 the company had moved from Trafford Park, Manchester, to Dagenham, Essex, and to avoid confusion with USA-built vehicles all English production carried an E suffix in the designation. Examples of T, TT, AF and BB are preserved.

Gilford

Built: 1926-38

Bodywork types: Single-deck bus and coach, double-deck bus

Engines:

Lycoming	– 6-cyl	5.984	petrol (166OT, 168OT 168SD)
Meadows	– 6-cyl		petrol (168MOT)
Buda	– 6-cyl	4.545	petrol (AS6)
Vulcan	– 6-cyl	7.4	petrol (176S, Hera, Zeus)
Dorman	– 6-cyl	6.9	oil (176S, Hera)
Leyland	– 6-cyl	8.6	oil (176S, Hera)
Tangye	– 6-cyl		2-stroke, oil (176S, Hera)

Wheelbase lengths: 14ft 0in (AS6)
16ft 8in (166OT, 168OT, 168SD)
17ft 6in (176S, Hera, Zeus)

Typical bodywork makes:
Eaton, Thurgood, Weymann, Willowbrook, Wycombe

Recognition features:
Wide upward and inward tapering radiator with Gilford name on top tank. Distinctive vertical Gruss air-springs on 168OT and 168SD each side of the radiator. The AS6 onwards had a narrower radiator with Gilford name on the mesh. Distinctively swept-back and down mudguards.

General remarks:
The Gilford Motor Co originated in E. B. Horne of London, N7 which specialised in importing products of the Garford Truck Co, Lima, Ohio. Gilford was formed in 1926 and in 1927 established premises at High Wycombe. In 1928 the 15OT and 166OT forward

Below:
A reminder of former glories at Victoria Coach Station in 1964 as this superbly restored Gilford 168OT set out for Brighton on the 10th British Coach Rally. Dating from 1931 and with Weymann C30D bodywork it has the earlier style Gilford radiator, high-set with prominent nameplate, and prominent Gruss vertical air springs. A. Hustwitt

Above:
At Brighton in 1968 at the end of the Historic Commercial Vehicle Run from London was this Eaton B20F-bodied Gilford AS6 also built in 1931. For the AS6 the radiator was modified to a slimmer form with oval top and bottom and with the Gilford name on the mesh.

control models were introduced and in 1929 the CP6. In 1929 the 168OT in forward and normal control versions was also announced together with the normal control 168SD. In 1930 the 20-seat AS6 appeared and the 168OT was offered with a British Meadows engine. Previous to this only the American Lycoming engine had been used. The Zeus double-decker arrived in 1932 followed in 1933 by its single-deck counterpart the Hera. Later that year the company moved to smaller premises at Park Royal and finally closed in 1938. During its final two years the title was Gilford (HSG) Ltd — the HSG standing for High Speed Gas. Vehicles were powered by producer gas, the only passenger example was a modified CF176 goods chassis with a converted AEC Regal engine bored out from 7.7 to 8.1 litres. It was sold to Cardiff Corporation which promptly converted it to diesel propulsion! Now preserved are examples of the AS6 (three), 166OT (one), 168OT (two), 168SD (two) and the chassis from a Hera.

Guy Early Types

Built: 1919-33
Bodywork types: Double- and single-deck bus
Engines: Knight – 4-cyl 35hp petrol
Guy – 4-cyl petrol
6-cyl petrol
Wheelbase lengths: A wide variety
Typical bodywork makes:
Brush, Dodson, Duple, Grose, Guy, Hall Lewis, Metro-Cammell, Ransomes, Strachans, Vickers
Variants: B, BA, BB, BKX, C, CX, FBB, FBKX, FBX, FC, OND, ONDF, ONDL, OWD Star Flyer
Recognition features:
Earliest examples had a very squat radiator with the Guy name on the top tank. Forward control models had a slightly tapering bonnet. The OND series introduced a more

distinctly tapered bonnet with vertically louvred side-panels. Springs protrude well forward of the radiator and headlamps.

General remarks:

In 1919 Guy goods chassis were made available for passenger applications but in 1921 it was decided to concentrate solely on the commercial vehicle market, a decision which led, in 1924, to the first true Guy passenger chassis. This was the BA, B and BB range, for 20, 26 and 30 seats respectively, among the first to use a dropped frame, allowing for a lower entrance and floor. In 1926 these were modified into the FBB of forward control design for up to 35 seats. At the same time a new 20-seater, the OWD, was introduced. The BX and BKX of 1926 were the first six-wheelers, the FBX and FBKX being their forward control equivalents. These were further improved in 1928 with the normal control CX and forward control FCX, of which the four-wheeled versions were classified C and FC. The FC was known as the Conquest (singe-deck) and Invincible (double-deck). Also in 1928 the OND and ONDF for up to 20-seat bodywork appeared and replaced the earlier model of that size, the BA. In 1931 the Star Motor Co was acquired and its famous Star Flyer continued in production for a further year when the OND was stretched to become the ONDL Victory. At the same time the four-cylinder engines were replaced by a new six-cylinder unit.

Preserved examples include a 1927 Vickers B32R-bodied FBB (registered YF 714) and a Grose-bodied double-deck FCX (NH 9189). Also possibly surviving are a Runabout (CD 9456) dating from 1917 and a 20-seat BA chassis (RR 3116) of 1926.

Guy Arab

Built: 1933-42 (Mk I and II), 1942-48 (Mk III), 1946-53 (Mk IV), 1951-62 (Mk V), 1962-68, 1950-59 (UF and LUF)

Bodywork types: Double- and single-deck bus, single-deck coach

Engines:

Gardner –	4-cyl	5.6	4LW type oil
	5-cyl	7.0	5LW type oil
	6-cyl	8.4	6LW type oil
	6-cyl	10.45	6LX type oil
Meadows –	6-cyl	10.35	oil

Wheelbase lengths: 16ft 3in (all to Mk III)
16ft 4in (Mk IV, Mk V, UF and LUF)
17ft 6in (single-deck Mk III)

Typical bodywork makes:

Brush, English Electric, Massey, Northern Counties, Park Royal, Pickering, Roe, Strachans, Weymann

Variants: Prewar — FD32 (32-seat single-deck), FD35 (35-seat single-deck), FD48 (56-seat double-deck) and FDX60 (six-wheel export double-deck)

Recognition features:

Tall radiator with oval inset at top bearing the Guy name. Mk Is with 6LW engines and Mk IIs onward have upswept front mudguards. Mk IVs were mostly fitted with concealed radiators, similar to those on later Daimlers. Some Mk IVs and Vs have 'Johannesburg fronts', namely a square grille with rounded corners having a double-tapering horizontal bar across the mid-point. An Indian's head often adorned the radiator cap on exposed radiator chassis.

General remarks:

The Arab provided a rugged, unsophisticated chassis, which placed the company among the country's leading bus manufacturers. In 1940 the Ministry of Supply contracted Guy Motors to produce a double-decker to utility specification, the result being the Arab Mk I of which 500 were built. Most of these were fitted with the Gardner 5LW engine but a few, such as Lincoln's, had the 6LW. The Mk II soon followed, mostly using the 6LW, and over 2,000 were built during the war years. After the war it continued in production until 1948, but those with the new lower radiator were known as Mk III and appeared from 1947. In 1946 a single-deck Mk III chassis appeared and additionally

Above:
Grimsby-Cleethorpes Transport 70 (JV 8641) was new to Grimsby Corporation in December 1943 and fitted with Park Royal utility bodywork with 56 seats. It was a Guy Arab Mk II 5LW and survived in service until 1962. This view shows it towards the end of its working life passing Grimsby Parish Church.

Below:
Lincoln City Transport 2 (DFE 385) was a Guy Arab III 6LW purchased in December 1948 and also fitted with Park Royal 56-seat bodywork though of the immediate postwar style. It lasted in service until 1966 when it was scrapped. This view shows it outside Lincoln Central station between trips on the city tour. Similar bus 23 (DFE 383) is now preserved with the Lincolnshire Vintage Vehicle Society.

Above:
Familiar to large numbers of local people this Guy Arab IV, formerly Birmingham Corporation 2548 (JOJ 548), was photographed in Hall Green in 1981. Features include the 'tin-front' concealed radiator, straight staircase and Metro-Cammell 54-seat bodywork. Mike Wood

Below:
Now among the ranks of preserved buses Huddersfield Joint Omnibus Services No 1 (FVH 1) was an all Guy Motors production — an Arab UF with 43-seat bus bodywork built in May 1951. It now resides with the East Pennine Transport Group.
Ian Allan Library

offered the Meadows engine. This chassis did not have the upswept front mudguards. In 1950 a Mk III of Southampton Corporation (No 162) was the first bus to be fitted with Cave-Brown-Cave heating. The Arab UF with underfloor engine was introduced in the same year and used the Gardner HLW series engines. A lighter version, the LUF, came in 1952. Meanwhile plans were made for a Mk IV double-deck chassis which went into production in 1951 with an export single-deck version available from the following year. The final Arab, the Mk V, arrived in 1962, providing for a forward entrance, low-floor bus in 27ft 0in and 30ft 0in options. Production of Guy vehicles finally ceased in 1968 following the takeover by Leyland, the last example bearing Leyland badges. Examples of each type of Arab have been preserved.

Guy Single-Deckers

Built: 1933-60

Bodywork type: Bus and coach

Engines:

Guy	4-cyl	3.7	petrol (Otter, Vixen from 1948)
Gardner	4-cyl	3.8	4LX type oil (Otter)
Meadows	4-cyl	22.5hp	EL type petrol (Wolf, prewar Vixen)
	4-cyl	5.43	4HDC330 type oil (Warrior LUF)
Perkins	4-cyl	3.15	P4 type oil (Wolf from 1952)
	6-cyl	4.73	P6 type oil (NLLVP and Seal)

Wheelbase lengths: 12ft 6in (Wolf and Vixen)
13ft 0in (Wolf from 1946)
14ft 9in (Vixen from 1947, Otter, Seal)
15ft 0in (NLLVP)
16ft 4in (Warrior LUF)

Typical bodywork makes:
Guy (until 1953), Alexander, Metalcraft, Thurgood, Eastern Coach Works (NLLVP), Mulliner (Douglas Corporation Otter)

Recognition features:
Early examples have sloping radiators with Guy name on top tank. Guy bodywork has flared skirting. NLLVP and Douglas Corporation Otters have the Red Indian motif on radiator top and rounded bonnets.

General remarks:
The Wolf CF14 and CF20 were normal control models introduced in 1933, the numbers indicating the seating capacity. In 1934 the Vixen 24 and 26 appeared as slightly larger versions. All were fitted with the Meadows four-cylinder petrol engine. After a wartime production break both types were reintroduced in 1947, the Wolf as a 20-seat normal control and the Vixen as a 30-seat forward control bus. Both remained available until 1956 and had redesigned radiators similar to that of the Arab. The range was updated in 1950 by the introduction of the 30-seat Otter, similar to but heavier than the Vixen and with the Gardner 4LK oil engine as an option. From 1952 the Perkins P6 was also available. In 1953-54, 84 examples of the famous London Transport GS-type were produced as a special order on the NLLVP chassis based on the Vixen but to LTE specification; all carried 26-seat ECW bodywork and almost half those built have been preserved. Superficially similar were five Mulliner-bodied Otters (Type NLLODP) delivered to Douglas Corporation in 1957. Many forward control Otters were built for export. In 1956 the Warrior straight-framed chassis for export was introduced and soon followed by the Warrior LUF for home sales, although the Arab LUF continued in production. In 1959 a short-lived derivative of the Vixen, the Seal, went into production and a small number found sales overseas. Preserved examples include the Wolf, Otter, Vixen and NLLVP.

Below left:
This 1938 Guy Wolf of Llandudno District Council saw many years service taking tourists around the Great Orme. Its Waveney bodywork seated 20 in coach seats and here in Rochdale its passengers were enjoying its comfort on the 1973 Trans-Pennine Run. There are no identification problems with the Red Indian Chief motif and the name on the radiator mesh. C. L. Caddy

Top:
Slightly larger and in the employ of Aberconwy District Council — Llandudno's successor — this Guy Otter was waiting for custom on the Great Orme tour. Bodywork is B25F by Charles Roe of Leeds and the bus was built in 1954 — 21 years before this view. It has now passed into preservation with similar CCC 597. K. Lane

Above:
The Guy NLLVP with ECW B26F bodywork, of which 84 were supplied to London Transport for rural services, has been popular with preservationists, no fewer than 30 having survived including GS26 (MXX 326) seen here at London Country's Hemel Hempstead open day in 1982. In spite of its age — it was built in 1953 — it is still a remarkably modern-looking design.

GRIMETHORPE
995
46
GUY
UCX 275
BOURNEMOUTH
MOBILE MUSEUM

UCX 275 is one of three surviving Guy Wulfrunians. It was new in 1961, with Roe 75-seat front entrance bodywork, as West Riding Automobile Services 995. It then became West Yorkshire PTE 405 before passing to preservationists who have fully restored it to its West Riding condition. D. M. Stuttard

Guy Wulfrunian

Built: 1959-65

Bodywork type: Double-deck bus

Engines:

Gardner –	6-cyl	8.4	6LW type oil
	6-cyl	10.45	6LX type oil
Leyland –	6-cyl	9.8	O.600 type oil

Wheelbase lengths: 15ft 4in
16ft 2in
17ft 0in
18ft 0in

Typical bodywork makes:
East Lancs, Northern Counties, Roe

Recognition features:
Full-fronted design with divided windscreen. Most had louvred front panels to ventilate the front-mounted engine but some East Lancs examples have Johannesburg-style fronts as fitted to some latter Arab Mk IVs and Vs. Air intakes are fitted on the front between-decks panelling. Front entrance examples have a nearside staircase.

General remarks:
The Wulfrunian was unveiled at the 1959 Commercial Motor Show and heralded as an advanced concept of double-deck bus design. Its overall height was only 13ft 3½in with normal height ceilings and central upper-deck gangway. The engine was mounted at the front alongside the driver but still allowing for a front entrance. Disc brakes, air suspension and a nearside staircase also featured. The offset front engine was connected to the drop-centre, double-reduction rear-axle via a diagonally running drive line and a gearbox between the axles. The design was too late to save Guy's difficult financial position and was too complex for most customers; and only 137 were built before the company's final demise within British Leyland Motor Holdings in 1965, Guy Motors having become bankrupt in 1960. Three front entrance Roe-bodied Wulfrunians have been preserved.

Below:
This was a fairly typical Karrier WL6 except for the unusual additional short upper deck which the operator advertised as the 'Tudec Super Safety Saloon'. The name of the operator is not known but the bodywork was by Strachan & Brown. Fortunately one WL6 has survived to be preserved — a former Ashton-under-Lyne English Electric-bodied saloon of 1928. Strachan & Brown

Above:
Several of the little Karrier 32A have now been preserved and, although not itself preserved, UXA 831 of 1958 is fairly typical with its Reading of Portsmouth 14-seat bodywork. Ian Allan Library

Karrier

Built: 1926-30 and 1952-64

Bodywork types: Single-deck bus and coach

Engines: Dorman – 6-cyl 6.96 JUL type, petrol (prewar models)
Commer – 4-cyl 2.666 petrol (postwar models)

Wheelbase lengths: 17ft 6in and 19ft 0in (WL6)
10ft 3in (32A)

Typical bodywork makes:
English Electric (WL6), Reading (32A)

Recognition features:
The WL6 is a three-axle single-deck model with forward entrance bodywork. The wide, square radiator has the headlamps mounted in front of it and the destination box is mounted under the front roof overhang alongside the driver's cab. The 32A is also a forward control chassis but built to carry bodywork of only 18ft 0in long by 6ft 9in wide. Distinct slope to the full-front and typical Rootes Group radiator grille and headlamps. Bumpers front and rear.

General remarks:
Karrier Motors Ltd, of Huddersfield, introduced the CL6 and WL6 models in 1926, two years before the first trolleybuses for which the company is better known. The six-wheel WL6 had both rear axles driven through a Karrier four-speed crash gearbox and worm-drive. Rear wheel and transmission brakes were fitted. The longer wheelbase version was known as the WL6/1 and both were produced until 1930 when, in the passenger market, the company began to concentrate on trolleybus production. Some 160 WL6 and WL6/1 models were built.

In 1952 the 32A was introduced as a small 14-seater for bus or coach application. Fitted with a four-speed synchromesh gearbox and hydraulic brakes it remained in production until 1964. From 1960 a Rootes four-cylinder 2.266-litre oil engine was offered as an alternative to the petrol version.

Leyland Early Types

Built: 1900-25

Bodywork types: Double- and single-deck bus

Engines: Leyland – 4-cyl 24-48hp petrol, but mostly rated 36-40hp

Wheelbase lengths: 11ft 9in to 16ft 6in
15ft 6½in (N)
15ft 10in (G7)
16ft 6in (SG7)

Typical bodywork makes:
Dodson, Leyland, Short Bros

Recognition features:
Low radiators with long bonnets on N and G7 normal control types. Leyland name on top tank of radiator, high, sheet-metal mudguards, thick tyres, crank-handle permanently fixed. The SG7, fitted with solid tyres, is halfcab, forward control, and the high-set radiator displays the Leyland badge.

General remarks:
The Lancashire Steam Motor Co was founded in 1898 in the village of Leyland and began to build lawnmowers and steam wagons. The first steam bus was constructed in 1900. In 1904 the first petrol-engined chassis was produced being, like all early chassis, suitable for goods or passenger application. As a result a large order was secured from the London & Suburban Omnibus Co in the following year. In 1907 the company was renamed Leyland Motors Ltd and two new chassis were introduced — the X and U types — both suited to goods or passenger use. The Leyland X should not be confused with that of the LGOC produced at its Vanguard Works in 1909. In 1912 the War Office opted for a 3-ton Leyland chassis as being suited to its needs, this becoming known as the RAF Chassis, 6,000 being produced for the War Office between 1914 and 1918. A major project after the war was the reconditioning of some 3,000 of these at a former aircraft factory at Kingston-upon-Thames. The N-type, similar in appearance to the RAF, soon appeared and remained available until 1923. A wide range of both normal and forward control chassis soon became available, of which the most popular was the G-type, particularly the G7 and SG7 (the S standing for Side-Type, ie the driver positioned alongside the engine). A lighter version was produced for the London market and designated LB, a double-decker with Leyland bodywork. A smaller normal control chassis marketed from 1920 was the single-deck B-type. By 1924 the company was well established as a major producer of bus chassis with the added facility of providing its own well designed bodywork, thus paving the way for those types which, in the ensuing years, were to make Leyland a household name.

Above right:
The beautifully restored Todmorden 14 (C 2367) has become a deservedly popular rally exhibit. A Leyland G2 it dates from 1921 and has opentop bodywork with 43 seats and is in every way a classic of its time.

Right:
The forward control Leyland G-type chassis was designated SG and this SG11 of Lee & Beaulah Services of Elloughton, East Yorkshire had B36R Leyland bodywork. No 11 (BT 8424) was sold out of service in April 1926 when only 11 months old and ended its days as a mobile workshop with Western Welsh. Sadly it was not preserved.
Leyland Museum

No 14
GARAGE ONLY
LEYLAND
HACKNEY CARRIAGE
C-2367
SPEED 12MPH

HULL & WITHERNSEA
LEE & BEAULAH SERVICES

Leyland Lion, Lioness, Cub and Cheetah

Built: 1926-40 (Lion), 1926-34 (Lioness), 1931-40 (Cub), 1935-40 (Cheetah)

Bodywork types: Single-deck bus and coach

Engines: Leyland – 4-cyl 3.96 petrol (Lioness LC1)
6-cyl 4.4 petrol (Cub KP2, KP3, KP4, SKP2, SKP3, SKP5)
6-cyl 4.4 oil (Cub KPO2, KPO3, SKPO2, SKPO3, SKPO5)
6-cyl 4.7 petrol or oil (Cub KPZ1, KPZ2, SKPZ2, KPZ3, KPZ4, Cheetah — all models)
6-cyl 4.7 oil (Cub REC)
4-cyl 5.1 petrol (Lion LSC1, PLSC3, LT1-3, LT5, LT5A, Lioness LC1, PLC1)
4-cyl 5.7 petrol (Lion LT5B)
4-cyl 5.7 oil (Lion LT5A, LT6-9)
6-cyl 6.8 oil (Lioness LTB1)
6-cyl 8.6 oil (Lion LT5B)

Wheelbase lengths: 13ft 0in½ (Cub KP4)
14ft 0in (Cub KP2, SKP2, KPO2, SKP02, KPZ1, KPZ3)
14ft 6in (Lion LSC1)
14ft 9in (Cub REC)
15ft 4½in (Lioness LC1)
15ft 6in (Cub KP3, SKP3, KPO3, SKPO3, KPZ2, SKPZ2, KPZ4, REC, Cheetah LZ3, LZ4)
16ft 3in (Cub SKP5, SKPO5)
16ft 5in (Lion PLSC3)
16ft 6in (Lion LT2)
16ft 7¼in (Lion LT1)
16ft 9in (Cheetah LZ1, LZ1A)
17ft 0in (Lioness PLC1, LTB1)
17ft 6in (Lion LT3, LT5, LT9)
17ft 7in (Lion LT5A, LT5B, LT6, LT7, Cheetah LZ2, LZ2A, LZ5)
17ft 8½in (Lion LT8)

Typical bodywork makes:
Alexander, Brush, Burlingham, English Electric, Harrington, Leyland, Massey, Park Royal, Roe, Thurgood, Weymann, Willowbrook. LPTB rear-engined Cubs (REC) — all LPTB

Recognition features:
The early Lions have very squat radiators with tapering sides. Later Lions, and Cheetahs, have a neater radiator design of slimmer and lighter appearance. Both are forward control designs. The Lioness and Cub are normal control models with tapering bonnets, the Lioness having the squat radiator of the Lion and the Cub a much smaller and narrower style. The Leyland name appears prominently on all models, often with the type name in addition.

General remarks:
The double-deck Leviathan and single-deck Leopard were the first models in Leyland's new L-range introduced in 1926 but it was the single-deck Lion which proved an instant success and enjoyed a production run into World War 2. From the original LSC1 model the Lion was progressively improved and found a wide application in both bus and coach use. Although not enjoying the success of the forward control Lion the normal control Lioness introduced in the same year sold in encouraging numbers, its 26-seat capacity being popular with tour operators. The lightweight Cub, built at the Kingston-upon-Thames factory from 1931, filled a gap at the small bus end of the market. Various models were available for up to 20-seat bodywork and from 1932 a forward control version (S prefix) for up to 30-seats was added. Production of Cubs ran until 1940 and the chassis proved extremely successful among rural operators with companies such as Lincolnshire Road Car taking large numbers. In 1935-36 Leyland successfully broke the AEC monopoly by selling Cubs to London Transport. These comprised 97 KPO3s and eight SKPZ2s, the latter having 1½-deck bodywork for the inter-station service. The success of these led to an order for 49 rear-engined versions (RECs) in 1939 to facilitate the introduction of one-person operation on more lightly used routes. The Cheetah of 1935 was essentially a lightweight version of the Lion and found popularity among the SMT fleets in Scotland.

Left:
Three Leyland Lions at the 1983 Crich Transport Extravaganza — in the centre TF 818 (Lancashire United 202), a Roe B30F-bodied LT1 of 1930, is flanked by a pair of 1927 B31F Leyland-bodied PLSC1s (KW 474 restored as Lincoln No 1 and CK 3825 which was Ribble 295). The PLSC1s have the original style of radiator and the LT1 the slimmer design which considerably improved appearances. M. Fowler

Top:
The Lion's radiator was improved still further with the introduction of the LT5A in 1934 as can be seen on this Alexander B36R example which originally saw service with Central SMT. In immaculate condition it was being rallied at a Strathclyde open day.
S. J. Brown

Variants

Lion

Marque	*Produced*	*Gearbox*	*Brakes*
LSC1	1926-27	Sliding mesh	Triple servo vacuum
PLSC3	1926-29	Sliding mesh	Triple servo vacuum
LT1	1929-30	Sliding mesh	Triple servo vacuum
LT2	1930-31	Sliding mesh	Triple servo vacuum
LT3	1931-32	Sliding mesh	Triple servo vacuum
LT5	1932-34	Sliding mesh	Triple servo vacuum
LT5A	1934-35	Constant mesh	Triple servo vacuum
LT5B	1934	Constant mesh	Triple servo vacuum
LT6	1934-35	Constant mesh	Triple servo vacuum
LT7	1935-37	Constant mesh	Triple servo vacuum
LT8	1938-40	Constant mesh	Triple servo vacuum
LT9	1938-40	Constant mesh	Triple servo vacuum

Note: The LT5B was supplied only to W. Alexander

Lioness

Marque	*Produced*	*Gearbox*	*Brakes*
LC1	1926-27	Sliding mesh	Triple servo vacuum
PLC1	1927-29	Sliding mesh	Triple servo vacuum
LTB1	1929-34	Sliding mesh	Triple servo vacuum

Cub

Marque	*Produced*	*Gearbox*	*Brakes*
KP2	1931-36	Constant mesh	Lockheed hydraulic
KP3	1931-36	Constant mesh	Lockheed hydraulic
KP4	1932-34	Constant mesh	Lockheed hydraulic
SKP2	1932	Constant mesh	Lockheed hydraulic
SKP3	1932-35	Constant mesh	Lockheed hydraulic
SKP5	1934-35	Constant mesh	Lockheed hydraulic
KPO2	1934-35	Constant mesh	Lockheed hydraulic
KPO3	1934-35	Constant mesh	Lockheed hydraulic
SKPO2	1934-35	Constant mesh	Lockheed hydraulic
SKPO3	1934-35	Constant mesh	Lockheed hydraulic
SKPO5	1934-35	Constant mesh	Lockheed hydraulic
KPZ1	1936-39	Constant mesh	Lockheed hydraulic
KPZ2	1936-39	Constant mesh	Lockheed hydraulic
SKPZ2	1936-37	Constant mesh	Lockheed hydraulic
KPZ3	1938	Constant mesh	Lockheed hydraulic
KPZ4	1938	Constant mesh	Lockheed hydraulic
REC	1938-40	Constant mesh	Lockheed hydraulic

Note: The REC was supplied only to London Transport (LPTB)

Cheetah

Marque	*Produced*	*Gearbox*	*Brakes*
LZ1	1935-36	Constant mesh	Hydraulic
LZ2	1935-37	Constant mesh	Hydraulic
LZ1A	1937	Constant mesh	Vacuum servo hydraulic
LZ2A	1937-38	Constant mesh	Vacuum servo hydraulic
LZ3	1937-38	Constant mesh	Vacuum servo hydraulic
LZ4	1938-40	Constant mesh	Vacuum servo hydraulic
LZ5	1938-40	Constant mesh	Vacuum servo hydraulic

Above:
Prewar Leylands usually carried the name of the chassis on the radiator mesh as on this Cub of Lincolnshire Road Car seen at Grantham Bus Station in 1933. LRCC operated over 100 Cubs of which this one (No 201 [FW 2779]) was a Brush B20F-bodied KP2 new in March 1932 and withdrawn in 1950. G. H. F. Atkins

Below:
The Leyland Cheetah was a lightweight Lion which saw only limited sales. Lincolnshire 586 (BFU 225), described by the company's records as a Leyland Light 6 4/SL, is thought to be a Cheetah LZ5 and was new in 1941 with Leyland DP35F body. It is seen here at the old Newark bus station when new. G. H. F. Atkins

Leyland Tiger, Tigress, Royal Tiger Worldmaster, Tiger Cub and Leopard

Built: 1927-39/45-54 (Tiger), 1934-39 (Tigress), 1950-54 (Royal Tiger), 1952-69 (Tiger Cub), 1955-79 (Royal Tiger Worldmaster), 1959-65 (Leopard L1 and L2)

Bodywork types: Single-deck bus and coach

Engines: Leyland –

6-cyl	6.8	petrol (Tiger TS1, TS2, TS3)
6-cyl	7.6	petrol (Tiger TS4, TS6, TS7, Tigress LTB3)
6-cyl	8.8	petrol (Tiger TS7)
6-cyl	7.4	oil (Tiger PS1)
6-cyl	8.6	oil (Tiger TS6, TS7, TS8, TS11, FEC, OPS1, Tigress LTB3)
6-cyl	5.76	O.350 type oil (Tiger Cub)
6-cyl	6.17	O.375 type oil (Tiger Cub from 1959)
6-cyl	6.54	O.400 type oil (Tiger Cub from 1963)
6-cyl	9.8	O.600 type oil (Tiger PS2, OPS2, OPS3, OPS4, Royal Tiger, Leopard L1 and L2)
6-cyl	11.1	O.680 type oil (Royal Tiger Worldmaster)

Wheelbase lengths:
15ft 7in (Royal Tiger PSU1)
16ft 2in (Tiger Cub PSUC1, Leopard L1, L2)
16ft 6in (Tiger TS3)
17ft 6in (Tiger TS1, TS2, TS4-9, TS11, PS1, PS2, OPS2, FEC, Tigress LTB3)
18ft 0in (Royal Tiger Worldmaster ERT1A)
18ft 9in (Tiger PS1/3, PS2/10-15)
19ft 0in (Tigress LTB3)
20ft 0in (Tigress LTB3, Royal Tiger Worldmaster ERT2A)
22ft 0in (Tigress LTB3)

Typical bodywork makes: Alexander, Brush, Burlingham, CIE, Crossley, Duple, E. Lancs, Eastern Coach Works, Harrington, Leyland, Massey, Metro-Cammell, Park Royal, Roe, Saunders Roe, Weymann, Whitson, Windover. LPTB Tiger FEC: all LPTB

Recognition features:

The Tiger TS1 and TS2 used a similar radiator shell to that of the Lion, but the TS3 saw this revised to a slimmer design with parallel sides. The Leyland badge is prominent on all radiator toptanks. The Tiger is of forward control design with a halfcab, the others having full-fronts. Most full-fronted examples carry the Leyland name on the front panel and this usually incorporates a representation of the animal after which the chassis is named.

General remarks:

In 1927 Leyland introduced the T-type six-cylinder 6.8-litre engine which, in turn, led directly to the production of the legendary Tiger and Titan range. The larger engine combined with a low floor height ensured that operators turned rapidly to the new designs. The TS series Tiger was progressively improved until production ceased with

Above right:
The first Leyland Tigers were similar in general appearance to the Lion and SO 3740 (Alexander P63), a TS2 with Alexander B32F body is typical of its breed. Paisley open day was host in July 1985. Maxwell M. Fowler

Right:
The design of the Tiger was soon modified and the postwar PS series chassis with coach bodywork was a classic of good design. ENT 778, a PS1 of 1948, bodied by Burlingham for 33 seated passengers was at the Showbus 1987. Hewitt of Watford (Premier-Albanian Coaches) is the present owner.

ALEXANDERS
GLASGOW
LIMITED STOP
P63
LEYLAND
SO 3740

PRIVATE
PREMIER
PREMIER
LEYLAND
PREMIER
ENT 778

the outbreak of World War 2. With a return to peacetime conditions production began immediately with the PS range with oil engines as standard and available for export (OPS series) as well as the home market. The Royal Tiger was introduced in 1950 as an underfloor-engined single-deck chassis for those operators unhappy with the integral Leyland/MCW Olympic. This was followed in 1952 by the lightweight Tiger Cub which, with successive improvements, enjoyed a 17-year production life. In 1954 the Royal Tiger Worldmaster was introduced for overseas customers and was basically a 'beefed-up' Royal Tiger with a larger engine and stronger chassis frame. It was an outstanding success and can still be found operating in many parts of the world, the last being built as recently as 1979. Many Worldmasters were built abroad under licence from Leyland. The Leopard of 1959 heralded a long-running series of underfloor-engined coaches and the L1 (bus) and L2 (coach) were essentially higher-powered Tiger Cubs. Their success led to the PSU3 series of 1961 and the subsequent PSU4 and PSU5 models.

Variants

Tiger

Marque	*Produced*	*Gearbox*	*Brakes*
TS1	1927-32	Sliding mesh	Triple servo vacuum
TS2	1928-32	Sliding mesh	Triple servo vacuum
TS3	1929-30	Sliding mesh	Triple servo vacuum
TS4	1931-33	Sliding mesh	Triple servo vacuum
TS6	1933-35	Constant mesh	Vacuum servo hydraulic
TS6T	1934-35	Constant mesh	Vacuum servo hydraulic
TS7	1935-37	Constant mesh	Vacuum servo hydraulic
TS7T	1935-37	Constant mesh	Vacuum servo hydraulic
TS7D	1935-37	Constant mesh	Vacuum servo hydraulic
TS8	1937-41	Constant mesh	Vacuum servo hydraulic
TS8T	1937-39	Constant mesh	Vacuum servo hydraulic
TS8D	1937-39	Constant mesh	Vacuum servo hydraulic
TS11	1941-42	Constant mesh	Triple servo vacuum
FEC	1939-40	Preselector	Air

Note: Torque converters were available as options as TS4 to TS8 and indicated by a 'c' suffix

Marque	*Produced*	*Gearbox*	*Brakes*	
PS1	1946-50	Constant mesh	Triple servo vacuum	Bus
PS1/1	1946-50	Constant mesh	Triple servo vacuum	Coach
PS1/3	1950-51	Constant mesh	Triple servo vacuum	Coach
PS1/4	1950	Constant mesh	Triple servo vacuum	
OPS1	1946-49	Constant mesh	Triple servo vacuum	Bus
OPS1/1	1946-49	Constant mesh	Triple servo vacuum	Coach
PS2/1	1948-51	Synchromesh	Triple servo vacuum	Bus
PS2/3	1948-51	Synchromesh	Triple servo vacuum	Coach
PS2/5	1948-51	Synchromesh	Triple servo vacuum	Bus
PS2/7	1948-51	Synchromesh	Triple servo vacuum	Coach
PS2/10	1947	Synchromesh	Triple servo vacuum	Bus
PS2/11	1947	Synchromesh	Triple servo vacuum	Bus
PS2/12	1950-54	Synchromesh	Triple servo vacuum	Bus
PS2/13	1950-54	Synchromesh	Triple servo vacuum	Coach
PS2/14	1950-54	Synchromesh	Triple servo vacuum	Bus
PS2/15	1950-54	Synchromesh	Triple servo vacuum	Coach

Note: PS1/4 was a double-deck model for Birch Bros
PS2/12 and 13 were originally wheelbase extensions of PS2/5 and 7, later denoted by an 'a' suffix

Below:
The Royal Tiger Worldmaster sold in huge numbers around the world and can still be seen in many places. Usually, but not always, identifiable by its badge, it carries bodywork produced in the country of operation. 666 291 belongs to the Israeli bus operator Egged and has Israeli Ha'argaz bodywork with wooden seats. Built around 1972 it was parked up at Jaffa Gate bus station in Jerusalem in February 1988 whilst working city services.

Marque	*Produced*	*Gearbox*	*Brakes*	
OPS2/1	1947-51			Bus
OPS2/3	1947-51			Coach
OPS3/1	1948-54		Triple servo vacuum	Bus
OPS3/3	1948-54	Synchromesh		Coach
OPS4/1	1948-54			Bus
OPS4/2	1954-58		Air	Coach, 'tin front'
OPS4/3	1948-54		Triple servo vacuum	Coach
OPS4/4	1954-58			Coach, 'tin front'
OPS4/5	1954-68	Pneumocyclic	Air	Bus
OPS4A/5	1967-68			'tin front'

Note: OPS2/1 used 7/32in framing
OPS3/1 used 9/32in framing
OPS4 models used 5/16in framing

Tigress

Marque	*Produced*	*Gearbox*	*Brakes*
LTB3	1934-39	Constant mesh	Vacuum servo

Royal Tiger (UK models)

All variants have synchromesh gearboxes

Marque	*Produced*	*Brakes*	
PSU1/9	1950-54	Triple servo vacuum	Bus
PSU1/10	1950-54	Air	Bus
PSU1/11	1950-54	Triple servo vacuum	Coach
PSU1/12	1950-54	Air	Coach
PSU1/13	1950-54	Triple servo vacuum	Bus
PSU1/14	1950-54	Air	Bus
PSU1/15	1950-54	Triple servo vacuum	Coach
PSU1/16	1950-54	Air	Coach
PSU1/17	1951-54	Triple servo vacuum	Leyland bus body

Royal Tiger (export models)

All variants have synchromesh gearboxes and air braking

Marque	*Produced*		
OPSU1/1	1950-55	17ft 6in wheelbase	Bus
OPSU1/2	1951-55	17ft 6in wheelbase	Coach
OPSU2/1	1950-55	20ft 4in wheelbase	Bus
OPSU2/2	1951-55	20ft 4in wheelbase	Coach
OPSU3/1	1950-55	15ft 7in wheelbase	Bus
OPSU3/3	1950-55	15ft 7in wheelbase	Bus

Tiger Cub (UK models)

All variants have air braking

Marque	*Produced*	*Gearbox*	
PSUC1/1	1952-62	Constant mesh	Bus
PSUC1/2	1953-62	Constant mesh	Coach
PSUC1/3	1953-62	Pneumocyclic	Bus
PSUC1/4	1955-62	Pneumocyclic	Bus
PSUC1/5	1955-62	Constant mesh	Bus
PSUC1/11	1962-69	Constant or synchromesh	Bus
PSUC1/12	1962-69	Constant or synchromesh	Coach
PSUC1/13	1955-62	Pneumocyclic	Bus
PSUC1A.13	1967-69	Modified pneumocyclic	Bus

Note: O.350 engine standard on PSUC1/1-5 inclusive from 1958, O.350 optional until 1961

Tiger Cub (export models)

All variants have air braking

Marque	*Produced*	*Gearbox*	
OPSUC1/1	1953-62	Constant mesh	Bus
OPSUC1/2	1953-62	Pneumocyclic	Bus
OPSUC1/3	1953-62	Pneumocyclic	Bus
OPSUC1/4	1953-62	Constant mesh	Bus
OPSUC1/11	1962-69	Constant mesh or synchromesh	
OPSUC1/12	1962-67	Pneumocyclic	
OPSUC1A.12	1968-69	Modified pneumocyclic	

Right:

Tiger Cubs with various body makes have been preserved but not East Yorkshire 670 (VKH 670) seen at Scarborough. Identity is by the badge on the dash-panel and bodywork is by Park Royal with 39 seats.

Royal Tiger Worldmaster

All models have pneumocyclic gearboxes and air brakes

Marque	*Produced*	*Wheelbase*	*Length*	*Width*
ERT1/1	1955-67	18ft 0in	34ft 0in	8ft 0in
CRT1/1	1955-67	18ft 0in	34ft 0in	8ft 0in
ERT2/1	1955-67	20ft 0in	36ft 0in	8ft 0in
CRT2/1	1955-67	20ft 0in	36ft 0in	8ft 0in
RT3/1	1955-67	16ft 3in	30ft 0in	8ft 0in
RT3/2	1955-67	16ft 3in	30ft 0in	8ft 0in
ERT1A.1	1968-73	20ft 0in	34ft 0in	8ft 0in
CRT1A.1	1968-73	20ft 0in	34ft 0in	8ft 0in
ERT2A.1	1968-73	20ft 0in	36ft 0in	8ft 0in
CRT2A.1	1968-73	20ft 0in	36ft 0in	8ft 0in
RT3A.1	1968-73	16ft 3in	30ft 0in	8ft 0in
ERT1B.1	1974-79	18ft 0in	34ft 0in	8ft 0in
CRT1B.1	1974-79	18ft 0in	34ft 0in	8ft 0in
ERT2B.1	1974-79	20ft 0in	36ft 0in	8ft 0in
CRT2B.1	1974-79	20ft 0in	36ft 0in	8ft 0in
RT3B.1	1974-79	16ft 3in	30ft 0in	8ft 0in

Note: All left-hand drive models prefixed 'L'

Leopard

Marque	*Produced*	*Gearbox*	*Brakes*	
L1	1959-65	Synchromesh	Air	Bus
LR2	1959-65	Synchromesh	Air	Coach

Above:
PJX 35 is a Leyland Leopard L1 with Weymann B42F body dating from 1962 but carries nothing to identify it — a not uncommon problem with full-front single-deckers. It was formerly Halifax Corporation 35 and is now preserved in full Halifax livery as seen in this Halifax open day scene in June 1987. D. J. Smithies

Leyland Titan

Built: 1927-42/45-69

Bodywork types: Double-deck bus and coach

Engines: Leyland – 6-cyl 6.8 petrol (TD1)
6-cyl 7.6 petrol (TD2, TD3, TD4 option)
6-cyl 7.9 petrol (TD5 option)
6-cyl 8.1 petrol (TD2)
6-cyl 7.4 oil (PD1)
6-cyl 8.6 oil (TD3-7, OPD1, OPD1A)
6-cyl 9.8 O.600 type oil (PD2, OPD2, PD3)

Variants: see separate Tables

Typical bodywork makes:

Prewar: Alexander, Beadle, Brush, East Lancs, English Electric, Leyland, Massey, Metro-Cammell, Park Royal, Strachans, Weymann, Willowbrook

Postwar: as above plus Burlingham, Duple, Metro-Cammell-Weymann, Northern Counties, Roe. Many Irish examples bodied by CIE; London Transport RTW bodied by Leyland, RTL by MCW, Park Royal and Weymann

---

Above right:
An original Titan TD1 now preserved is Bolton Corporation 53 (WH 1553) of 1929. It has open-staired 51-seat lowbridge bodywork and is in the care of the Lincolnshire Vintage Vehicle Society. This view shows it on an outing to the Newark Traction Engine Rally in 1972.

Right:
The Titan TD series was progressively modified until, by the TD5 of 1937, a very modern and attractive bus was available. This Lincolnshire Road Car TD5 of 1939 in the company's attractive brown and white livery was photographed at Skegness soon after delivery. It has Leyland all-metal 56-seat bodywork which has a thoroughly modern look about it, progress thus far set the scene for successful postwar developments. LRCC 567 survived until 1957 when it was exported to Yugoslavia. G. H. F. Atkins

Recognition features:
The TD1 and TD2 have a rather heavy design of radiator with slightly tapering sides and oval top. As on all Leylands the name appears on the radiator toptank. The windscreen has a distinct rake and early bodywork was often of the 'piano-front' design. Later TD series vehicles have a lighter design of radiator with parallel sides and thinner frame. The PD series radiator is shorter with a slightly curved front. The 'tin front' design is a full-width bonnet with vertical slots whilst the 'St Helens front' is a glass-fibre moulding first introduced to the order of St Helens Corporation in 1960; this has the Leyland name at the midpoint.

General remarks:
The legendary Titan is probably the bus for which Leyland is best known. Introduced in 1927 it was produced continuously (apart from a two-year wartime break) until 1969. The TD1 of 1927 was a revolutionary double-deck design and one in which the AEC Regent of 1929 found a hard rival. The Titan and Regal were to become by far the most popular British front engined double-deckers. The Titan, when fitted with the standard Leyland lowbridge body, was only 13ft 0in high and led to a rapid resurgence of double-deck use outside the capital. TD type variants continued to be built into World War 2 and from 1933 were available as 'gearless buses', fitted with torque converters — models suffixed 'c', eg TD3c. In 1937 Leyland won an order for 100 TD4s from London Transport, the last 10 of which were 'gearless'. In 1945 the new PD1 made its appearance and was followed two years later by the PD2 of which over 2,100 were supplied to London Transport as its RTL and RTW classes. Legislation in 1956 allowing for 30ft 0in-long double-deckers led to the PD3 range and both PD2 and PD3 variants continued to be built up to the end of production in 1968. The PD2 was also available for export from 1948 and in this form was prefixed 'O'. A large number of Titan variants was available depending on the operator's choice of engine, gearbox, brakes, radiator style, chassis length and bodywork size, etc. As might be expected a very large number of Titans — around 500 — has now been preserved.

Below:
A significant postwar version of the Titan PD2 was developed for London Transport in 7ft 6in (RTL) and 8ft 0in (RTW) widths. Bodywork of the RTL was interchangeable with the AEC Regent IIIRT. RTW 497 (LLU 987) was new in 1950 and has Leyland 56-seat all-metal bodywork to the standard LT design.

Variants

Marque	*Produced*	*Wheelbase*	*Gearbox*	*Brakes*	*Radiator*
TD1	1927-31	16ft 6in	Sliding mesh	Triple servo vacuum	Exposed
TD2	1932-33	16ft 6in	Sliding mesh	Triple servo vacuum	Exposed
TD3	1933-35	16ft 6in	Constant mesh	Triple servo vacuum	Exposed
TD4	1935-37	16ft 6in	Constant mesh	Servo hydraulic	Exposed
TD5	1937-41	16ft 6in	Constant mesh	Servo hydraulic	Exposed
TD6c	1938-39	16ft 3in	Torque convertor	Servo hydraulic	Exposed
TD7	1939-42	16ft 6in	Constant mesh	Triple servo vacuum	Exposed

Note: The torque convertor was available as an option on TD3, TD4, TD5 and TD7 denoted by 'c' suffix

PD1	1945-48	16ft 3in	Constant mesh	Triple servo vacuum	Exposed
PD1A	1946-51	16ft 3in	Constant mesh	Triple servo vacuum	Exposed
PD1/1	1947	16ft 3in	Constant mesh	Triple servo vacuum	Exposed
PD1/2	1947	16ft 3in	Constant mesh	Air	Exposed
PD1/3	1947-49	16ft 3in	Constant mesh	Triple servo vacuum	Exposed
OPD1	1946-48	17ft 6in	Constant mesh	Triple servo vacuum	Exposed
OPD1A	1946-48	17ft 6in	Constant mesh	Triple servo vacuum	Exposed

Note: 'A' suffix denotes Metalastik rubber bushes on road spring shackle pins
PD1/1 supplied only to the City Coach Co
PD1/2 supplied only to Bolton Corporation
PD1/3 8ft-wide variant

PD2/1	1947-51	16ft 3in	Synchromesh	Triple servo vacuum	Exposed
PD2/3	1947-51	16ft 3in	Synchromesh	Triple servo vacuum	Exposed
PD2/4	1948-49	16ft 3in	Synchromesh	Air	Exposed
PD2/5	1948-50	16ft 3in	Synchromesh	Air	Exposed
PD2/9	1954	16ft 5in	Synchromesh	Triple servo vacuum	Exposed
PD2/10	1950-56	16ft 5in	Synchromesh	Triple servo vacuum	Exposed
PD2/11	1954-55	16ft 5in	Pneumocyclic	Air	Exposed
PD2/12	1950-57	16ft 5in	Synchromesh	Triple servo vacuum	Exposed
PD2/13	1950-56	16ft 5in	Synchromesh	Air	Exposed
PD2/14	1953-54	16ft 5in	Preselector	Air	Exposed
PD2/15	1951	16ft 5in	Pneumocyclic	Air	Exposed
PD2/20	1953-57	16ft 5in	Synchromesh	Triple servo vacuum	Tin front
PD2/21	1953-57	16ft 5in	Synchromesh	Air	Tin front
PD2/22	1953-57	16ft 5in	Synchromesh	Triple servo vacuum	Tin front
PD2/23	1953-57	16ft 5in	Synchromesh	Air	Tin front
PD2/24	1954-59	16ft 5in	Pneumocyclic	Air	Tin front
PD2A/24	1960-67	16ft 5in	Pneumocyclic	Air	St Helens
PD2/25	1954-59	16ft 5in	Pneumocyclic	Air	Tin front
PD2A/25	1960-62	16ft 5in	Pneumocyclic	Air	St Helens
PD2/26	1954-58	16ft 5in	Pneumocyclic	Air	Lightweight Tin front
PD2/27	1955-59	16ft 5in	Synchromesh	Air	Tin front
PD2A/27	1960-67	16ft 5in	Synchromesh	Air	St Helens
PD2/28	1955-59	16ft 5in	Synchromesh	Air	Tin front
PD2A/28	1960-61	16ft 5in	Synchromesh	Air	St Helens
PD2/30	1955-59	16ft 5in	Synchromesh	Triple servo vacuum	Tin front
PD2A/30	1960-66	16ft 5in	Synchromesh	Triple servo vacuum	St Helens
PD2/31	1955-59	16ft 5in	Synchromesh	Triple servo vacuum	Tin front
PD2A/31	1960-62	16ft 5in	Synchromesh	Triple servo vacuum	St Helens
PD2/34	1958	16ft 5in	Pneumocyclic	Air	Exposed
PD2/35	1956-62	16ft 5in	Pneumocyclic	Air	Exposed
PD2/37	1956-66	16ft 5in	Synchromesh	Air	Exposed
PD2/38	1956-62	16ft 5in	Synchromesh	Air	Exposed
PD2/40	1956-66	16ft 5in	Synchromesh	Triple servo vacuum	Exposed

Marque	*Produced*	*Wheelbase*	*Gearbox*	*Brakes*	*Radiator*
PD2/41	1956-62	16ft 5in	Synchromesh	Triple servo vacuum	Exposed
PD2A/44	1967-68	16ft 5in	Pneumocyclic	Air	St Helens
PD2/47	1967-68	16ft 5in	Synchromesh	Air	St Helens
PD2A/54	1967-68	16ft 5in	Pneumocyclic	Air	Exposed
PD2/57	1967-68	16ft 5in	Synchromesh	Air	Exposed
6RT	1948-51	16ft 4in	Preselector	Air	Exposed
7RT	1948-54	16ft 4in	Preselector	Air	Exposed

Note: PD2/4 supplied only to Bury and Bolton Corporations
PD2/5 supplied only to Blackpool Corporation
PD2/9 supplied only to St Helens Corporation
PD2/11 supplied only to Leeds Corporation
PD2/15 only one prototype constructed
PD2/34 supplied only to Manchester Corporation
6RT supplied only to London Transport (RTW class)
7RT supplied only to London Transport (RTL class)
Many PD2/1s supplied with constant mesh gearbox

OPD2/1	1948-54	17ft 6in	Synchromesh	Triple servo vacuum	Exposed
OPD2/2	1948-54	17ft 6in	Synchromesh	Air	Exposed
OPD2/6	1954	17ft 6in	Synchromesh	Air	Tin front
OPD2/7	1955-58	17ft 6in	Pneumocyclic	Air	Tin front
OPD2/9	1956	17ft 6in	Pneumocyclic	Air	Exposed
OPD2/10	1956	17ft 6in	Synchromesh	Air	Exposed
PD3/1	1956-60	18ft 6in	Synchromesh	Air	Tin front
PD3A/1	1960-66	18ft 6in	Synchromesh	Air	St Helens
PD3/2	1956-60	18ft 6in	Pneumocyclic	Air	Tin front
PD3A/2	1960-66	18ft 6in	Pneumocyclic	Air	St Helens
PD3/3	1956-60	18ft 6in	Synchromesh	Triple servo vacuum	Tin front
PD3A/3	1960-62	18ft 6in	Synchromesh	Triple servo vacuum	St Helens
PD3/4	1956-67	18ft 6in	Synchromesh	Air	Exposed
PD3/5	1956-66	18ft 6in	Pneumocyclic	Air	Exposed
PD3/6	1956-66	18ft 6in	Synchromesh	Triple servo vacuum	Exposed
PD3/11	1967-68	18ft 6in	Synchromesh	Air	St Helens
PD3A/12	1967-68	18ft 6in	Pneumocyclic	Air	St Helens
PD3/14	1967-69	18ft 6in	Synchromesh	Air	Exposed
PD3A/15	1967-68	18ft 6in	Pneumocyclic	Air	St Helens

Above right:
The 'tin-front' concealed radiator was popular with a number of operators and Lincoln Corporation took six of these Titan PD2/31s with Roe 61-seat bodies in 1957 which survived until the mid-1970s. No 75 (KVL 679) is seen near the city centre in the earlier green and white livery which later gave way to all-over green, Fortunately No 80 (KVL 684) lasted long enough to become a training vehicle and then to be preserved.

Right:
OIK 984 is a Titan OPD2/10 built for export in 1958 and supplied to CIE in Ireland. It was bodied in the CIE workshops and seats 68 passengers. Now preserved it was seen whilst touring the UK mainland early in 1988. Notice the curve of the radiator shell and the autovac on the front bulkhead.

4
DODDINGTON RD
KEEP LEFT
75
KVL 679

PICKO TEA The Best
PIGEON HOUSE
1
PEARSE STREET
GEVAERT
FILM
CLUB LEMON
R788
OIK 984

Above:
The Titan PD3 was commonly fitted with a St Helens front as seen in this view of ex-Leicester 73 (CJF 73C) seen in service with Weardale Motor Services in 1983. Bodywork for 74 passengers is by East Lancs and the vehicle dates from 1965.

Leyland Miscellaneous Chassis

Built: 1925-27 (Leviathan), 1926 (Leopard), 1926-28 (Leveret), 1927-39 (Titanic), 1930-36 (Badger), 1937/9/40 (Gnu and Panda), 1948-71 (Comet), 1949-71 (Olympic)

Bodywork types: Double-deck bus (Leviathan and Titanic)
Single-deck bus and coach (remainder)

Engines: Leyland –

4-cyl	3.96	petrol (Leveret LA2, Badger BA4)
4-cyl	5.1	petrol (Badger BA4)
6-cyl	6.5	petrol (Leviathan LSP2, Leopard LSG2)
6-cyl	6.8	petrol (Titanic TT1)
6-cyl	7.6	petrol (Titanic TT2)
6-cyl	8.0	petrol (Leviathan LG1, LSP1)
6-cyl	8.6	oil (Gnu, Panda, Titanic TT3-5)
6-cyl		O.300 oil and P.300 petrol (Comet CPO1 and CPP1)
6-cyl	5.76	O.350 type oil (Comet ECPO2)
6-cyl	6.17	O.370 type oil (Comet 12C.6)
6-cyl	6.54	O.400 type oil (Comet 13C.6)
6-cyl	6.54	O.401 type oil (Comet 13C.16)
6-cyl	9.8	O.600 type oil (Olympic HR series, Leopard L1-2)
6-cyl	11.1	O.680 type oil (Olympic Series II and III)

Typical bodywork makes: Olympic — MCW integral; Comet — Duple, Harrington, Park Royal; Others — Burlingham, Duple, Harrington, Leyland, Plaxton, Roe, Thurgood, United A.S.

Variants: see following Tables

Recognition features:
The Leviathan and Titanic were double-deck models with the wide, squat radiator as used on the Lion and Titan TD1, the Titanic being a three-axle design. The Leveret was a normal control model using the narrower radiator of the Lioness, the Badger was similar but with a longer chassis. The Gnu and Panda were three-axle single-deckers with twin-steering. The Comet was a normal control design with distinctive curved and fluted bonnet which curved out to meet the mudguards which also incorporated the headlamps. Its radiator grille consisted of horizontal chromed slats. The Olympic was an integral forward control, full-fronted single-decker based on the Tiger Cub and looking almost identical to it.

General remarks:
The double-deck Leviathan was the first of Leyland's new L range introduced in 1925. In two years of production less than 100 chassis were sold and the design showed little advance on earlier designs. The Leopard of 1926 was an inconspicuous single-deck model soon displaced by the popularity of the Lion. Also in 1926 came the Leveret, a normal control chassis for 20-seat bodywork. Its withdrawal in 1928 left a gap in the Leyland small bus range which was not filled until the Cub was introduced in 1931. A radical step in 1927 was the introduction of the three-axle Titanic, an unfortunate name for a model no one seemed to want. Its failure to compete against the AEC Renown led to only small numbers being built although it remained available until 1939. During the gap between Leveret and Cub a number of operators specified passenger bodywork on the Badger goods chassis, a normal control design which was suitable for up to 20 seats. Leyland's final prewar designs were the three-axle Gnu and Panda, both using twin-steering front axles. The Gnu had a front-mounted engine whilst the Panda's was underfloor. The use of three axles allowed bodywork up to 30ft 0in long. Only eight Gnus and one Panda were built. The year 1948 saw the introduction of the semi-forward control Comet as a middleweight chassis for home and export markets. Whilst the UK take-up did not last long the model did well overseas and continued to be available until 1971, the same year in which the Olympic ceased production. This had been introduced in 1949 in conjunction with MCW as an integral underfloor engined alternative to the Tiger, but operators' preference for separate chassis/body combinations led to its comparative lack of success. For the export market the Olympic was progressively

CEDRIC
RACES
REV 329
LNK 786

RAMSEY
R9
47
Road Services
PAY
AS YOU
ENTER
Road Services
47
MMN 297

improved and orders continued until the end of its availability. Examples of the Badger, Comet and Olympic are now preserved.

Variants

Leviathan

Marque	*Built*	*Wheelbase*	*Gearbox*	*Brakes*	*Notes*
LG1	1925-26	15ft 11¼in	Sliding mesh		
LSP1	1926-27	15ft 11¼in	Sliding mesh		
LSP2	1927	15ft 11¼in	Sliding mesh		

Leopard

Marque	*Built*	*Wheelbase*	*Gearbox*	*Brakes*	*Notes*
LSG2	1926	16ft 5in	Sliding mesh		Only two built

Leveret

Marque	*Built*	*Wheelbase*	*Gearbox*	*Brakes*	*Notes*
LA2	1926-28	12ft 9in	Sliding mesh		

Titanic

Marque	*Built*	*Wheelbase*	*Gearbox*	*Brakes*	*Notes*
TT1	1927-31	17ft 1½in	Sliding mesh		
TT2	1932-34	17ft 2in	Sliding mesh		
TT3	1935-37	17ft 2in	Constant mesh		
TT4	1935-37	18ft 10½in	Constant mesh		
TT5	1937-39	17ft 1½in	Constant mesh		
TT6	1937-39	18ft 10½in	Constant mesh		Available but none built

Badger

Marque	*Built*	*Wheelbase*	*Gearbox*	*Brakes*	*Notes*
BA4	1930-36	14ft 6in	Sliding mesh	Mechanical	

Gnu

Marque	*Built*	*Wheelbase*	*Gearbox*	*Brakes*	*Notes*
TEP1	1937	16ft 3in	Constant mesh		Only three built
TEC2	1939	16ft 3in	Constant mesh		Only five built

Panda

Marque	*Built*	*Wheelbase*	*Gearbox*	*Brakes*	*Notes*
	1940	16ft 3in	Constant mesh		Only one built underfloor engine

---

Above left:
The Leyland Comet was a unique normal control design with a high bonnet of distinct profile. LNK 786 was built in 1949 for Albanian Coaches of St Albans later passing to Cedric of Wivenhoe and had coach-seated bodywork by Strachans. G. J. Wadley

Left:
Isle of Man Road Services 47 (MMN 297) was a Leyland HR40 Olympic integral with bodywork by MCW for 40 seats and was new in 1950. The Olympic was visually like the Tiger Cub and, apart from the badge, hard to identify. Similar bus 84 (MMN 302) has been saved for preservation. Ian Allan Library

Comet

Marque	*Built*	*Wheelbase*	*Gearbox*	*Brakes*	*Notes*
CPO1/ CPP1	1948-52	17ft 6in	Constant mesh	Vacuum hydraulic	Optional 15ft 10in wheelbase
ECP02.1	1952-62	16ft 11in		Vacuum hydraulic	
ECP02.2	1953-62	14ft 8in		Vacuum hydraulic	
ECP02.3	1956-60	16ft 11in		Vacuum hydraulic	Semi-forward control
ECP02.4	1956-60	17ft 6in		Vacuum hydraulic	Semi-forward control
12C.6RP	1962-67	16ft 11in		Air	
13C.6RP	1963-68	16ft 11in		Air	
13C.16RP	1969-71	16ft 11in		Air	

Note: 12C and 13C series — 'R' indicates right-hand drive, for left-hand drive versions 'L' is substituted, eg 12C.6LP

Olympic

Marque	*Built*	*Wheelbase*	*Gearbox*	*Brakes*	*Notes*
HR40	1949-50	15ft 7in	Synchromesh	Air or vacuum	Home market only
HR44/ 90/S	1950-53	15ft 7in	Synchromesh	Air or vacuum	Home market only
HR44/ 96/S	1950-53	15ft 7in	Synchromesh	Air or vacuum	Home market only
E36	1950-54	15ft 1in	Synchromesh	Air or vacuum	Export market only
E40	1950-54	17ft 6in	Synchromesh	Air or vacuum	Export market only
E44	1950-54	20ft 4in	Synchromesh	Air or vacuum	Export market only

Olympic Series II

Marque	*Built*	*Wheelbase*	*Gearbox*	*Brakes*	*Notes*
E (R or L) 2/40	1955-65	12ft 8in	Pneumocyclic	Air	
E (R or L) 2/44	1955-65	20ft 0in	Pneumocyclic	Air	

Note: 'L' and 'R' indicate left- and right-hand drive versions respectively

Olympic Series III

Marque	*Built*	*Wheelbase*	*Gearbox*	*Brakes*	*Notes*
EL3/41	1966-68	20ft 0in	Pneumocyclic	Air	
E (R or L) 3/44	1966-68	20ft 0in	Pneumocyclic	Air	
EL45/37	1968-70	20ft 0in	Pneumocyclic	Air	Istanbul order only
EL3A.41	1969-70	20ft 0in	Pneumocyclic	Air	Rationalised gearbox
ER3A.44	1969-71	20ft 0in	Pneumocyclic	Air	

Maudslay

Built: 1903-60

Bodywork types: Single-deck bus and coach

Engines: Maudslay – 4-cyl 4.9-litre petrol (ML3)
AEC – 6-cyl 7.7-litre oil (Marathon)

Wheelbase lengths: 16ft 6in (ML3 series)
17ft 7in (Marathon)

Typical bodywork makes: Burlingham, Northern Counties, Park Royal, Thurgood, Whitson

Recognition features:

The early chassis had normal control with a large radiator bearing the Maudslay name and mesh-vents in the side frames. They also had large solid-tyred, spoked wheels. The ML3 has a parallel-sided radiator with the Maudslay name on the toptank, the radiator side frames are wider at the centre than at the ends. The Marathon carries the typical AEC radiator but incorporating the Maudslay shield. Both are forward control, halfcab designs.

General remarks:

The Coventry-based company built its first passenger vehicles in 1903 based on its three-cylinder 20hp car chassis. The first proper bus appeared in 1904 as a six-seater on the same chassis. There then followed a large fleet delivered to the Great Western Railway comprising closed and charabanc types of 40hp. Until 1912 production comprised mostly buses. In 1924 the AEC/LGOC NS type chassis was used as a pattern for a new bus, the ML3, produced from the following year. Of forward control design power was provided by a four-cylinder petrol engine and four-speed gearbox. In 1929 a six-wheeled model was announced but found few purchasers. In 1933 the SF40 was introduced, having a front overhang allowing for a front entrance and 40 seats. In 1946 the Marathon III was introduced, being almost identical to the AEC Regal III and powered by the same engine. It remained in production until 1950 when it was replaced by the Marathon IIIA. In 1948 the company had combined with AEC and Crossley to form Associated Commercial Vehicles Ltd, models competing with AEC marques were soon being discontinued and the factory switched to components manufacture. By 1951 all true Maudslays had been withdrawn from manufacture and in 1953 the plant was closed. However, until 1960 certain AEC models could be purchased with the Maudslay badge. Examples of the ML3, ML3B and Marathon III are preserved.

Below:
The Maudslay ML3 chassis owed much to the AEC 4 series and had a similar but modified radiator design and 'proper' headlamps. YR 1089 is a good example of the type clearly showing these features. Bodywork was B32R by Buckingham and it was new as No 1202 to the Great Western Railway in September 1926, passing to Thames Valley Traction in June 1931. Ian Allan Library

Above:
Coventry Museums is the owner of this Maudslay Marathon III of 1948 with Trans-United C33F bodywork, originally with Hackets of Manchester. The chassis is almost identical to the AEC Regal III but with Maudslay's own design of radiator featuring a prominent company badge. It is seen here on the 1981 Trans-Pennine Run.
W. M. Ewing

Milnes-Daimler

Built: 1902-14

Bodywork types: Charabanc, double and single-deck bus

Engines: Daimler 4-cyl 16, 20, 24, 28 and 30hp petrol

Wheelbase length: 10ft 3½in

Bodywork makes: Milnes-Daimler and a great variety of local builders

Recognition features:
Large 'coal-scuttle' bonnet which replaced the earlier Renault style. Milnes-Daimler name on radiator toptank. Large canopy over driving position, solid tyres, very thin mudguards, very high chassis.

General remarks:
The Milnes-Daimler chassis, of German construction, came on to the British market in 1902, the Great Western Railway being one of the first purchasers of 16hp 22-seat examples with Milnes-Daimler bus bodies. By 1904 the company's bodyshop had been closed and customers had to look elsewhere to complete their chassis. In 1904 the first

examples appeared in London where, for the next few years, they were operated by the London Road Car Co Ltd, Thomas Tilling, Birch Bros and the London Motor Omnibus Co, the latter obtaining priority on Milnes-Daimler output. Most of the London buses were open-top 34-seat examples with outside stairs and powered by 24hp engines. From 1905 goods models were also produced but, with its dependence on imports from Germany, the company ceased trading at the outbreak of hostilities in 1914. Kent-registered D 1959 of 1903 is preserved in the USA and London-registered A 9164 is preserved in its home city.

Above:
Early Milnes-Daimler chassis had bonnet covers similar to the Renault, namely a forward-sloping top and a radiator set low down between the front wheels. This was soon replaced by a 'coal-scuttle' style as found on No 5 (LC 3311) of the London & North Western Railway. Oil lamps and sheet-metal mudguards were fitted and bodywork was 34-seat by Dodson. A 24hp model, it went into service in the Watford area in April 1906 and in 1913 the body was transferred to a Commer. LNWR/Ian Allan Library

Morris Commercial

Built: 1924-51

Bodywork types: Double and single-deck bus

Engines: Morris – 4-cyl 13.95hp petrol (1 Ton T)
6-cyl 25.01hp petrol (CS11/30, CV11/30 & 40, CVF13/5)
6-cyl 5.1 petrol (Imperial)
6-cyl 7.7 petrol (Dictator)

Wheelbase lengths: 10ft 2in (1 Ton T)
11ft 2in (CS11/30)
11ft 6in (CV11/30 & 40)
13ft 6in (CVF13/5)
16ft 4¾in and 18ft 0in (Dictator and Imperial)

Typical bodywork makes: Bracebridge, Metro-Cammell, Morris, Taylor, Wadham

Recognition features:
The 1 Ton T has a narrow tapering bonnet with Morris-Commercial name cast into the radiator toptank. Normal control design. The Dictator has a tall parallel-sided radiator with the Dictator name on the mesh. The C-series have a variety of radiator styles but with tapering bonnets and name on the radiator toptank.

General remarks:
Morris Commercial Cars Ltd represented W. R. Morris's move into the commercial light truck market in 1924 when he purchased the Soho (Birmingham) works of E. G. Wrigley & Company. First to appear was the 1 Ton T using the 1.8-litre engine fitted to Morris's cars. It was claimed to be the first mass-produced truck and found many applications as a small bus chassis. In 1929 the Dictator HF appeared as a forward control single-deck chassis which, in 1931 was lengthened by 1ft 8in. Its double-deck equivalent, the Imperial HD followed the same year and, with its offset drive-line allowed for a low floor and lowheight bodywork. Most of the Imperials built were bought by Birmingham Corporation. The semi-forward control C-series began production in 1933. Designed basically as a goods chassis they found many uses in the passenger market. The CS11/30 and CV11/30 were for 30cwt payloads and the CV11/40 for two tons. The CVF13/5 of 1937 was designed to carry 27-seat bodywork. Production of the 1 Ton T ceased in 1932, the Dictator and Imperial in 1935 and the C-series in 1951. Two 1 Ton Ts are preserved together with a Dictator, a CS11/30 and a CV11/40.

Below:
The Morris Dictator was a very attractive design as this pre-delivery picture shows. The Morris motif and Dictator name appear on the radiator and an autovac is mounted on the front bulkhead of the Short Bros rear entrance bodywork. The date is 14 November 1931. Short Bros

Northern General Transport

Built: 1933-39

Bodywork type: Single-deck bus and coach

Engines:

AEC	4-cyl		6.6 oil (SE4)
Hercules	6-cyl	WRXT type	6.30petrol (SE6 bus)
	6-cyl	WXLC3 type	6.65petrol (SE6 coach)

Wheelbase length: 16ft 7½in

Bodywork makes: Short Bros (SE6), English Electric (SE4)

Recognition features:
Full-fronted design, entrance ahead of leading axle, windscreen divided vertically in ratio of 1:2. Slight forward slope on the side window line at rear, front off-side window curving down to coincide with bottom edge of windscreen. Trapezoidal radiator mesh in front panel divided vertically in half. Single-track destination blind with a smaller route blind on each side.

General remarks:
In 1933 NGT designed and built an underfloor, side-engined, three-axle chassis 10m long to achieve the maximum possible seating capacity in a single-deck body. Of the two rear axles only the forward one was driven (via a Fuller, then a David Brown four-speed gearbox). The American Hercules engine was chosen as no British unit would fit the very restricted space available. Only the first six SE6s were built by NGT at its Bensham works, production being transferred in 1935 to AEC at Southall. In that year it was decided to build a two-axle 40-seat version and this was achieved by reducing the chassis length and rear body overhang. The resulting SE4 was of the then maximum permitted length of 27ft 6in. Thirteen SE4s were built at Bensham between 1936 and 1939, the remaining 12 being constructed at Southall. Some 42 SE6s of both 44-seat bus and 28-seat coach design were built between 1933 and 1936 of which a number were later converted to two-axles. The preserved SE6 (registration CN 6100) was built in 1934 as a three-axle bus but was converted to two-axles in 1943 and can now be seen at the North of England Open Air Museum, Beamish.

Below:
Short Bros built the body for this NGT SE6 photographed at Milton Street, Nottingham in July 1935. It has the standard NGT radiator grille and features of interest include the windscreen design, the open folding roof and the rake of the door and windows.
G. H. F. Atkins

This Morris CVF13/5 is shown in another pre-delivery view — taken on 3 May 1947 according to the information on the back of the print — at the Morris factory in Oxford. Of semi-forward control the bonnet is noticeably short, aligning well with the 27-seat bodywork. Morris Motors Ltd

REO

Built: 1924-33
Bodywork types: Single-deck bus and coach
Engines: REO 4-cyl 30hp and 6-cyl 27.3hp petrol
Wheelbase lengths: 10ft 8in (Speedwagon)
11ft 11in (Pullman Junior)
13ft 0in (Gold Crown)
14ft 7in (Pullman)
Bodywork makes (preserved examples): Harrington, Fordingbridge (Speedwagon)
Taylor (Pullman)
Economy (Gold Crown)

Recognition features:
All models are normal control. The Speedwagon has a parallel-sided, round-topped bonnet and radiator with high-set headlamps, thin sheet-metal mudguards and a double-strip front bumper, two-piece flat windscreen and triangular quarter-lights. The Pullman and Gold Crown have tapered, oval-topped bonnets and radiators, the radiator being protected by six vertical steel bars. All have swept mudguards and spoked wheels.

General remarks:
R. E. Olds set up the REO Motor Car Co at Lansing, Michigan in 1904 and the one-ton Speedwagon made its first appearance in 1915 using a four-cylinder petrol engine of 27hp with three-speed transmission. It was introduced to the British market in 1924 when a serious attempt was made to win business from the many small independent operators of the time. By 1925 many examples were using the six-cylinder engine and production ceased in 1928 when the Pullman, a much larger coach model was introduced. It was powered by the same engine and had seats for 26. It continued in production until 1933. In 1929 the Gold Crown FB became available as a direct replacement for the Speedwagon and this again used the six-cylinder engine but a four-speed crash gearbox. It remained in production until 1931. No further models were offered on the British market but REO continued to produce buses and trucks in the USA until 1967.

Below:
The famous REO Speedwagon was a popular design with many rural operators in the 1920s and 1930s and MR 3879 has fortunately survived. It was new to Harrington of Fordingbridge in 1925 and bodied by an unknown local builder with 14 coach seats and folding roof. Seen in Otley in 1977 it was taking part in the Manchester to Harrogate Run. M. Fowler

Above:
The Reo Pullman was larger than the Speedwagon and replaced it in 1928. This Taylor C23F-bodied example — NG 1109 — originally with Reynolds of Overstrand and new in 1931, is now preserved on the Isle of Wight. The radiator design is that used on both Pullman and Gold Crown models.

Rowe

Rowe's Garage Ltd of Liskeard was a coach operating and garage concern owned by M. G. Rowe who began vehicle building in 1953 with a prototype, front-engined, single-deck chassis called the Hillmaster. This used a Meadows four-cylinder 5.43 oil engine, five-speed Meadows gearbox and Moss rear axle. In 1954 an underfloor-engined version was introduced of which three examples were built using the same engine and transmission. The chassis was available until 1960 but only two further examples were constructed — in 1956 — of which one survives in the care of the North Cheshire Historic Vehicle Group. This has a 16ft 6in wheelbase and full-fronted bodywork by Reading of Portsmouth incorporating a front-entrance, two-piece windscreen and single-headlamps. Its registration is WRL 16. Rowe continued to produce truck chassis until 1962.

Seddon

Foster & Seddon of Salford began truck production in 1938 but was halted by the outbreak of hostilities in 1939. Production then resumed in 1947 when a new factory was opened at Oldham. In 1948 the company introduced its Mk IV passenger model as a forward control chassis of 14ft 11in wheelbase for 24ft 6in long single-deck bodywork. A Perkins six-cylinder 4.73-litre P6 type oil engine was used in conjunction with a five-speed constant mesh gearbox. Vacuum-servo hydraulic braking was standard. An improvement from 1955 was the fitting of a two-speed rear axle. Production ceased in 1963 and two examples are now preserved. Other passenger chassis of the period included the Mk XI of 1952, an underfloor-engined chassis using the same Perkins P6 power unit, and the Mk XIX of 1957 using an underfloor AEC six-cylinder 7.7-litre oil engine. The main recognition feature of Seddon vehicles is the emblem comprising a circle on which is super-imposed the title 'Seddon diesel'.

Above:
The price was £3,945 complete for this 1953 Seddon Mk XI underfloor-engined chassis with Duple Elizabethan C41C bodywork used as a demonstrator by both companies before taking up service with George Moore & Son of Lytham St Annes. A Perkins P6 diesel engine was used in this very attractive vehicle. Ian Allan Library

Sentinel

Built: 1949-54

Bodywork types: Single-deck bus and coach

Engine: Sentinel – 4-cyl 6.08 oil

Wheelbase lengths: 14ft 9in

Typical bodywork makes:
Sentinel (STC4), Beadle (SLC4)

Recognition features:
Full-front, forward control, both types with four-piece 'drooping' windscreens and roof-mounted destination indicators. Radiator grilles consist of horizontal chromed bars over a mesh screen radiused inwards from headlamp centres. The Sentinel name appears prominently on the front panel over the grille. Offside door provided for access by the driver.

General remarks:
The Sentinel Waggon Works Ltd of Shrewsbury is famous for its long production of steam road and rail vehicles. The company was formed in 1917 and built its first bus in 1939 upon its taking over Gilford Ltd. This was known as the Gilford HSG and reference is made to it under the Gilford heading. Sentinel's first conventional bus appeared in 1949 as the STC4, an integrally constructed single-decker 27ft 6in long and seating 40 passengers. In 1950 the SLC4 was introduced as a 35-seat coach with the same overall dimensions as the STC4 and powered by the same engine and transmission. It was not, however, of integral construction and all were fitted with bodywork by Beadle of Dartford. Production of the STC4 ended in 1953 and the SLC4 in 1954. During 1952 the company changed its name to Sentinel (Shrewsbury) Ltd and by the mid-1950s was concentrating entirely on railway vehicle production. The HSG, two STC4s and four SLC4s have been preserved.

Top:
Three Sentinel STC4 40-seat buses seen before delivery to their respective operators in 1951. The bodywork for all STC4s was to the same design and had this very distinctive concealed radiator and offside driver's door. Two examples are now preserved, though none of these. Ian Allan Library

Above:
The Beadle-bodied all-Sentinel SLC4 coaches had an alloy body giving an unladen weight of only 5½ tons. PHA 928 is typical of its breed and was built in early 1951 for Eagle Coaches of Smethwick (A. Grainger Ltd). Ian Allan Library

Shelvoke & Drewry

Shelvoke & Drewry of Letchworth introduced its small-wheeled Freighter goods model in 1923. Power was provided by a transverse-mounted four-cylinder 2.2-litre petrol engine and transmission was by a three-forward and three-reverse epicyclic gearbox driving a worm rear axle via bevel gearing. Control was by two tram-type controllers, the right-hand one for steering and the left-hand one for transmission and brakes. A transmission brake operated automatically depending upon the position of the controller. The chassis was available with a choice of 6ft 6in, 8ft 0in, 9ft 6in or 11ft 0in wheelbases. Solid tyres were fitted at first but in 1927 pneumatics became standard on new vehicles. In 1929 a 13ft 3in wheelbase was offered and a 2.4-litre engine fitted. A 3-litre engine became available in 1932 and wheel-steering in the following year. Production continued until 1938 by which time over 2,500 had been built although less than 50 of these were fitted with passenger bodywork.

Above:
Almost everything about the Shelvoke & Drewry Freighter was unusual as this picture of PX 886 operating in Worthing shows. Most notable are the tiny wheels and the radiator of the transverse-mounted engine.

Straker-Squire

Built: 1906-25

Bodywork types: Double- and single-deck bus

Engines: Büssing – 4-cyl 28.9hp petrol
4-cyl 32.8hp petrol

Wheelbase length: 13ft 6in

Typical bodywork makes:
Local bodybuilders

Recognition features:
Solid-tyred wheels, long bonnet with high radiator top tank, single headlamp mounted in front of radiator, high-set chassis (much higher than contemporary Leylands, etc).

General remarks:
The early Straker-Squire was the current German-built Büssing sold in the UK by Sidney Straker and Squire of Bristol. They were three-tonners with 28/30hp, four-cylinder petrol engines and supplied for both goods and passenger use. A three-speed gearbox was fitted which drove through a cone clutch. By 1908, 356 double-deckers were operating in London. Vehicles were increasingly assembled from British-made parts and the firm reorganised in 1913 and introduced a 3/4-tonner which saw extensive military use during World War 1. Despite a brief affair with trolleybuses (Straker-Clough) from 1921 to 1925 the company suffered in the postwar depression and went into liquidation in 1926. Two examples have been preserved, DL 621 is a chassis only dating from 1912 and WT 7101 (Keighley Corporation No 5) is a Straker-Clough trolleybus of 1924 carrying Brush bodywork on a 14ft 6in wheelbase.

Thornycroft

Built: 1907-57
Bodywork type: Double- and single-deck bus, single-deck coach, charabanc

---

Below:
Thornycroft BC MN 5942 was originally a demonstrator before becoming Isle of Man Road Services No 54. It was new in 1928 with a Strachan & Brown body and is seen languishing in Laxey Depot in 1959. Similar Hall Lewis-bodied MN 5454 has been preserved by IOM National Transport. A. Moyes

Bottom:
Also introduced in 1927 was the Thornycroft A2 normal control chassis for 20-seat bodywork of which this example was caught by the official photographer before sale. Like the BC, production of the A2 ceased in 1930. J. I. Thornycroft & Co Ltd

Engines: Thornycroft – 2-cyl 16hp petrol (earliest models and early J)
4-cyl 30hp, 40hp and 50hp petrol (later J)
4-cyl 22.5hp petrol (A2)
6-cyl 40.8hp petrol (BC)
6-cyl 46hp petrol (FC and HC)
6-cyl 29.4hp petrol (Lightning)
6-cyl 3.865 type ER4 petrol (Nippy)

Wheelbase lengths: 13ft 0in, 13ft 7½in, 14ft 4½in, 16ft 6in (J)
14ft 0in (A2)
16ft 6in (BC)
18ft 0in (FC and HC)
14ft 8in (Lightning)
12ft 6in (Nippy)

Typical bodywork makes:
Dodson, Hall Lewis, Harkness, Vickers, many local builders

Variants: Early models in 1½-, 2- and 2½-ton normal control versions, mostly bodied as charabancs. Others as indicated in tables above and text below.

Recognition features:
Early models had extremely 'heavy' design radiators with round tops and large top tanks, slightly tapered bonnets and solid tyres. The radiator was soon modified to an oval top and bottom design with a central, vertical dividing strip, a bar across the front and large headlamps mounted alongside. Later models have a higher radiator with slightly convex top plus a dividing strip. The Thornycroft name usually, though not always, appears on a small oval plate on the top tank. The Nippy has a large T on the radiator mesh.

General remarks:
John I. Thornycroft was originally a builder of steam launches who built his first road wagon in 1896 at his Chiswick factory. By 1907 he had turned to petrol-engined commercials, introducing his first passenger vehicle shortly afterwards. The J type was introduced in 1913 as a normal control model suitable for double- and single-deck bus and charabanc use. The chassis was progressively stretched and the engine size increased until the model was withdrawn in 1926. The A2 followed in 1927, this being a normal control chassis for 20-seat bodywork which continued in production until 1930. Also in 1927 came the BC, a forward control low-frame chassis for 28 to 32-seat bodywork. This was also built until 1930. The FC of 1930 was a three-axle forward control single-deck chassis for 40-seat bodywork and produced until 1932. The HC was its double-deck counterpart introduced in 1929. In 1931 came the 32-seat Cygnet single-decker and the Daring double-decker, and in 1933 the Lightning normal control single-decker was introduced which was produced until 1936. The Nippy HF appeared after World War 2 — in 1946 — and was a forward control goods chassis of which a number found bus applications. Production of this model continued until 1950. In 1948 the SG/NR6 coach made its appearance but met with little success. From then on production turned more and more to heavy trucks and few buses were built after 1950.

Left:
Liverpool Corporation Tramways was a buyer of the Thornycroft FC 32-seat single-decker. Designed for bodywork up to 40 seats the Liverpool examples sat 37 in rear-door bodywork built in the Tramway workshops. Real Photographs

Tilling-Stevens

Built: 1911-53

Bodywork types: Double- and single-deck bus, charabanc, single-deck coach

Engines:

Tilling-Stevens –	4-cyl	33.15hp	and 32.4hp petrol (TS series)
	4-cyl	39hp	petrol (B9 and 10 series)
	4-cyl	5.13	petrol (B10 and 49 series)
Gardner –	5-cyl	7.0	5LW type oil (K series)
	6-cyl	8.4	6LW type oil (K series)
Meadows –	6-cyl		6DC.630 type oil (K series)

Wheelbase lengths: 14ft 6in, 15ft 6in, 16ft 6in (TS series)
15ft 6in (B9A and B10A, B49A)
17ft 0in (B10C)
17ft 3in (K series)

Typical bodywork makes:
Beadle, Brush, Eastern Coach Works, Strachans, plus local builders

Recognition features:
Early models fitted with 'coal-scuttle' bonnets, name in large letters on top tank, solid tyres, high chassis frame. The more modern B series retained a heavy but modified radiator assembly having a large top tank and almost semi-circular bottom panel, the side panels being swept out to the support brackets, two-piece inclined windscreen. Some B series had vertical bars over the radiator. The K series has a parallel-sided radiator with a slightly dished top tank, the whole being curved back towards the top.

General remarks:
Thomas Tilling, famous as one of the London horsebus pioneers and forerunner of the National Bus Co, began production of his TTA1 chassis in 1911 as a joint venture with W. A. Stevens of Maidstone. The TTA1 was a petrol-electric model in which power was derived from a 20kW dc generator driving the rear axle through an electric motor. The TTA2 followed in 1913 as a 40hp chassis for both double- and single-deck bodywork. This was redesigned as the TS3 in 1914 and soon improved to become the TS4. Power was initially provided by a 36hp petrol engine directly coupled to a compound-wound generator supplying current to a series-wound electric motor driving a worm-geared rear-axle via a propeller shaft. No gearbox was fitted, speed and torque being controlled by resistances. The TS3A of 1921 was fitted with a 32.4hp engine. At the end of World War 1 the conventionally driven 4-ton TSB3 and 2½-ton TSB4 made their appearance, whilst, at the same time a series of petrol-electric models numbered TS1 to TS7 soon became available and remained in production until 1926. In that year the B series made its first appearance as a conventional-transmission forward control chassis using 29hp petrol engines. The B9 was built from 1926-31 and the B10 from 1927-31, the former being of straight-frame and the latter of dropped-frame construction. Braking was mechanical, acting on the rear wheels only but in 1928 servo-assisted braking became available on the B10A leading to the designation B10A2. A longer version, the B10C, was introduced in 1927. This had a dropped-frame design initially with mechanical braking but with servo-assisted braking after 1928 when it became the B10C2. Production ended in 1931. The following year saw the B49A7 of basically similar design with a production run which lasted only until 1933. At the end of 1947 Tilling-Stevens introduced its new Express K series chassis using Gardner or Meadows diesel engines — designations being K5LA7 (Gardner 5LW), K6LA7 (Gardner 6LW) and K6MA7 (Meadows 6DC.630). In all models transmission was via a five-speed constant mesh gearbox. Production ended in 1953. Examples of TTA2, TS1, TS3, TS6, B9A, B10A2, B10C2 and K6MA7 have been preserved.

MACCLESFIELD
CHEADLE
HANDFORTH
WILMSLOW
ALDERLEY
MACCLESFIELD
TIZER
CUTTY SARK
SCOTS WHISKY
TILLING STEVENS
DB 5070

SOUTHAMPTON
GOU 732

Left:
The early Tilling-Stevens chassis were distinguishable by the large symmetrical radiator carrying the words 'Tilling-Stevens Petrol-Electric'. DB 5070 is a TS6 of 1925 carrying replica Brush 54-seat opentop and stairs bodywork and actively preserved by the Chalk Pits Museum in Surrey. S. C. Morris

Below left:
The Tilling-Stevens postwar K series managed somehow to look a little old-fashioned. 1949 K6MA7 GOU 732 with Scottish Aviation C33F body was photographed at a Brighton rally. Recognition features include the parallel-sided radiator curving back, the autovac on the front bulkhead and the curves to bottom of front panel, radiator and mudguards.

Trojan

Trojan Ltd of Croydon, a subsidiary of the Trojan Truck Manufacturing Co of Los Angeles, began producing light commercials in 1949. The only passenger model was the DT built from 1953-60, one of which has been preserved. The DT was a normal control model having a 7ft 10in wheelbase suitable for bodywork up to 13 seats. A four-cylinder 2.36-litre Perkins P3 type oil engine was used in conjunction with a three-speed synchromesh gearbox, single-plate clutch and hydraulic brakes. Most examples were used as personnel carriers and only three are known to have been used as 'proper' buses. The DT has a narrow tapering bonnet, headlamps incorporated in the mudguards and a high four-piece windscreen with a rather sad, drooping look. The preserved example is fitted with Strachans bodywork.

Below:
OVB 345 is a delightful example of the little Trojan DT. Built in 1955 with 13-seat bodywork it was used as a personnel carrier by the Napier Aircraft Co before its preservation. D. Trevor Rowe

Vulcan

The Vulcan Motor & Engineering Co Ltd of Southport was a subsidiary of Driggs-Seabury Ordnance Co of Sharon, Pennsylvania which, from 1913, began to build its own versions of the British-designed Commer Car range. The Southport factory built a variety of passenger and goods models between 1914 and 1950. In 1931 the company went into receivership and was bought out by Tilling-Stevens in 1938, Tilling-Stevens being bought out by the Rootes Group in 1950. The VSC four-cylinder side-valve petrol-engined normal control bus chassis was introduced in 1925 and soon followed by the VSD. In 1928 came the 'Brisbane' chassis using a six-cylinder 6.6-litre engine in a drop-frame chassis. The Emperor double-deck model followed in 1930 and the Duchess single-deck soon after. The Duchess was a normal control single-deck chassis for up to 32 seats. One postwar passenger chassis was produced from 1949 and, using a Perkins six-cylinder 4.7-litre P6 type oil engine, it had a 15ft 3in wheelbase. It met with little success. A VSD and a Duchess are now preserved.

Below:
Two Vulcans have fortunately been saved for preservation, a VSD and a Duchess. Several VSDs were operated by Tom Carey at Skegness on a seafront service which began in 1926. They passed to Skegness Motor Services in 1932 and two years later to the Lincolnshire Road Car Co which operated them until 1952. FU 5946 had 25-seat open-sided bodywork by Eaton — seats were wooden, tyres were solid and the fare one penny which could be used, on production of the ticket, for a ride at Butlins funfair. That was in 1929. Notice the Vulcan name on the bonnet cover. G. H. F. Atkins

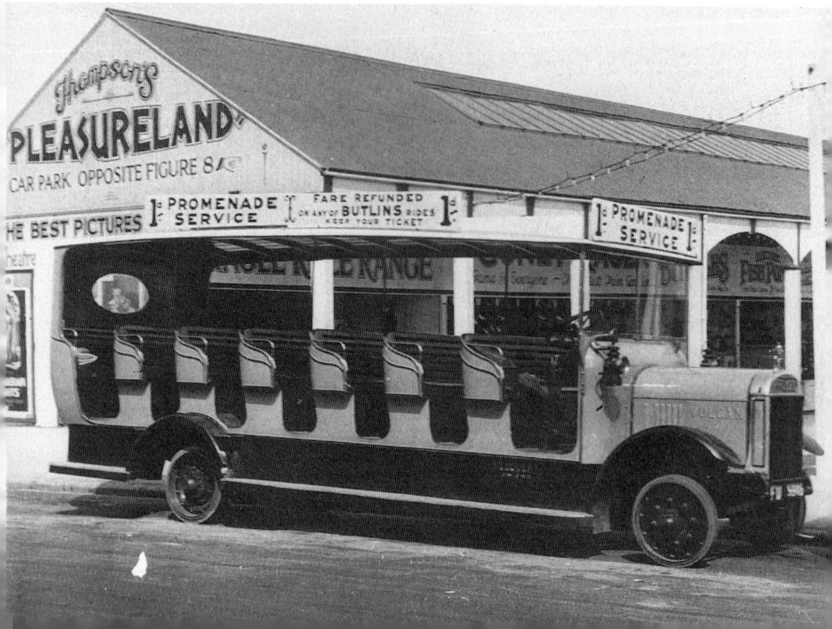

Appendix

Radiator Selection

AEC Regent IIIRT

Albion

AEC Regent V

Bedford OB

AEC Routemaster

Bedford — Duple Vega

Beadle-Leyland. S. C. Morris

BMMO SOS

BMMO S15. BMMO

Bristol — postwar K and L

Bristol Lodekka — first series

Bristol Lodekka — final series

Bristol SC coach

Bristol MW

Commer-Harrington

Crossley DD and SD. R. Garnett

Daimler exposed radiator

Daimler concealed radiator

Dennis postwar

Dennis Loline III

Guy

Guy NLLVP

Leyland TD1

Leyland PD3/1

Leyland badge